FRUITS OF PROGRESS

GROWING SUSTAINABLE FARMING AND FOOD SYSTEMS

LORI ANN THRUPP

WORLD RESOURCES INSTITUTE

APRIL 2002

ELAINE LIPSON
EDITOR

HYACINTH BILLINGS
PRODUCTION MANAGER

MAGGIE POWELL
LAYOUT

CLEMENS KALISCHER
COVER PHOTO

Each World Resources Institute report represents a timely, scholarly treatment of a subject of public concern. WRI takes responsibility for choosing the study topics and guaranteeing its authors and researchers freedom of inquiry. It also solicits and responds to the guidance of advisory panels and expert reviewers. Unless otherwise stated, however, all the interpretation and findings set forth in WRI publications are those of the authors.

Copyright © 2002 World Resources Institute. All rights reserved.

ISBN 1-56973-472-0

Library of Congress Control Number: 2002105588

Printed in the United States of America on chlorine-free paper with recycled content of 50%, 20% of which is post-consumer.

Images on pp. 61, 63, and 75 courtesy of Wine Institute of California

CONTENTS

Acknowledgments v

Executive Summary vii

PART I: Overview of Lessons Learned

Chapter 1.
Introduction to the Fruits of Progress 1

Chapter 2.
Characteristics and Motivations of
Sustainable Food and Agriculture
Enterprises 5

Chapter 3.
Main Ingredients and Strategies
for Greening the Food System 21

Chapter 4.
Effective Practices Used in Sustainable
Food Production and Marketing 31

Chapter 5.
Challenges and Actions
To Expand Progress 43

PART II: Case Studies

Del Cabo 56
Durst Farmers 58
Fetzer Vineyards 60
Frog's Leap 62
Full Belly Farm 64
Lagier Ranches 66
Lodi Woodbridge Winegrape Commission 68
Lundberg Family Farms 70
Natural Selection Foods 72
Robert Mondavi Winery 74
Sherman Thomas Ranch 76
Small Planet Foods 78

References 81

Appendix

People interviewed and
consulted for the study 85

ACKNOWLEDGMENTS

The creation of this report was made possible through the collaboration of many organizations and individuals. I am grateful to all of the innovative enterprises and people involved in the profiles of sustainable agriculture and food systems in this report. These cases and the main contacts for each include Del Cabo (Larry Jacobs), Durst Farms (James Durst), Fetzer Vineyards (Paul Dolan), Frogs Leap (John Williams, Full Belly Farm (Judith Redmond), Lagier Ranches (John Lagier), Lodi Woodbridge Winegrape Commission (Mark Chandler, President and Cliff Ohmart), Lundberg Family Farms (Bryce Lundberg), Natural Selection Foods (Myra and Drew Goodman, CEOs), Robert Mondavi Winery (Tim Mondavi and DeWitt Garlock), Sherman Thomas Ranch (Mike Braga), and Small Planet Foods (Gene Kahn, CEO). I appreciate the added assistance of David Runsten and Cliff Ohmart in providing background details for the cases of Del Cabo and Lodi Woodbridge Winegrape Commission respectively.

I thank Arthur Getz, Associate at World Resources Institute, for his valuable work, insights, and camaraderie throughout the entire project. I appreciate the excellent input, support and feedback of WRI reviewers and colleagues, Don Doering, Paul Faeth, Thomas Fox (formerly WRI Vice President), Tony Janetos, Nels Johnson, Tony LaVina, Patricia Londono, and Don Reed. I am also indebted to external reviewers, Jill Auburn, Jenny Broome, John Ikerd, Richard Kashmanian, Fred Kirchenmann, Karen Ross, and James Tischer, whose comments and suggestions were extremely helpful. Special thanks to many other people who provided insights, support, and/or information, including Michael Abelman, Miguel Altieri, Susan Clark, DeWitt Garlock, Cathy Greene, Bruce Hirsch, Bruce Jennings, Walter Knausenberger, Tim LaSalle, Mark Lipson, Craig McNamara, Stephen Pavich, Judith Redmond, Mark Ritchie, Bob Scowcroft, Sean Swezey, and many additional colleagues whose names I might have overlooked unintentionally.

We greatly appreciate the generous support and patience of the U.S. Environmental Protection Agency's Policy, Economics and Innovation Program, Swedish International Development Agency, the Great Valley Center, and the U.S. Agency for International Development, which helped to fund the project. I also am thankful to Elaine Lipson for her editorial expertise, Hyacinth Billings and Maggie Powell for valuable work on graphics, editing and production, and Cristina Balboa, Cecilia Blasco, and Anne Marie DeRose for program assistance on this project.

Last but not least, I thank the innovative growers and enterprises who are boldly cultivating sustainable approaches and developing socially-responsible "green" growth in the food and agriculture sector. They provide inspiration and hope for the global society, the economy, and for future generations.

Executive Summary

FRUITS OF PROGRESS: GREENING THE FOOD SYSTEM

A "green" transformation is sprouting in the food and agriculture industry. Growing numbers of farmers, food manufacturers, and distributors in many parts of the world are adopting environmental stewardship approaches and other methods to protect public health and natural resources. For a variety of economic, social, and environmental reasons, businesses are integrating ecological considerations into farming practices, food factory operations, and grocery shelves, just as individuals are addressing these concerns in their daily food selections. This report will demonstrate that this approach holds multi-faceted benefits for these businesses and for society.

Although organic and ecological farming budded during the late 1960s as a relatively small counter-culture movement in the United States and Europe, it has grown and changed dramatically since then, blossoming globally into a multi-billion-dollar mainstream business. This 'green' sector in the food and agriculture industry (including producers, manufacturers, and distributors) is now expanding at an unprecedented rate.

These innovators are responding strategically to rising consumer demand for foods that are produced in environmentally responsible or 'natural' ways. They are using environmental stewardship practices, and forging a new 'state of the art' in food systems, setting an important trend and leadership for the agriculture and food sector in the twenty-first century. Even large conventional foods corporations and venture capitalists are increasingly investing in the natural foods business, drawn by attractive market opportunities. This 'green' transition is spreading worldwide, with international implications for how foods are produced and marketed.

In this report, we've identified this remarkable growth of environmental stewardship in the food and agriculture industry as 'greening the food system.' 'Green' refers broadly to a range of approaches that are interpreted to be 'sustainable,' meaning methods that are environmentally sound as well as socially responsible and economically viable. The term "sustainable" farming may include certified organic practices, and also encompasses other ecological and integrated practices.

Fruits of Progress identifies the drivers behind the changes taking place, and some of the main elements and strategies for developing sustainable food and agriculture approaches. The report identifies salient common features of innovators involved in the green transformation, based largely on case studies. It shows how ecologically based practices can generate profits, while contributing to broader goals of sustainable development.

We also identify challenges and barriers to progress including lack of information, research and policy support for sustainable farming practices, and the growing concentration of the industry. We propose actions and identifying opportunities for continued growth of sustainable and organic food systems. The lessons and guidelines presented here are intended to be useful for decision-makers in the food and agriculture industry, and for policy-makers and government agencies that influence this industry. It is also relevant for economic analysts and consumers interested in food and farming issues.

GREEN GROWTH TRENDS

Although this report does not focus on the organic sector alone, the organic market offers a good illustration of this fast-paced change. During the 1990s, the certified organic food market grew very rapidly, at an annual average rate of about 20 percent internationally, and 25 percent in the United States. The growth rate of the more mature conventional food industry during this same time was less than 5 percent per year. The total global retail sales of organic foods was estimated at $21.5 billion in 2000 while sales in the U.S. organic market reached an estimated $7.8 billion in 2000, a 20 percent increase over 1999 figures.

Europeans have experienced the highest growth rate of organic production and marketing in the world. At the same time, North American and Japanese organic markets are rapidly catching up, and organic markets are gaining ground in developing countries as well. The organic sector is being transformed from a very small niche segment and a movement of mostly small farmers, to a mainstream industry. Experts expect this dynamic growth to continue in the future, especially with the advent of national organic standards for the United States, overseen by the National Organic Program of the U.S. Department of Agriculture.

In addition to the organic approach, a growing number of agricultural producers and manufacturers are using diverse environmental stewardship practices, ranging from soil conservation methods to integrated pest and crop management and recycling of materials, in some areas and crops. The adoption of these practices and the expansion of markets for organic and sustainably produced foods are likely to continue, as consumer demand grows and ecological innovations spread globally.

MAIN LESSONS FROM EXPERIENCES

We conducted case studies for this report on a group of diverse food and agriculture innovators that are developing sustainable and/or organic approaches. These are relatively well-known operations in the fruit and vegetable industry, based in the western United States - primarily California, where there has been remarkable progress in innovative ecological approaches to agriculture. These innovators in the case studies

> **BOX 1 | COMPELLING CAUSES BEHIND ADOPTING SUSTAINABLE APPROACHES**
>
> We identified the following factors as important driving forces for implementing sustainable practices:
>
> *Caring for the land.* Pioneers in sustainable farming have deep concerns about land stewardship and environmental responsibility, and strive to maintain the health of soil and resources.
>
> *Consumer demand for environmentally sound practices.* Public opinion about food increasingly impacts farmers' choices.
>
> *Competitive advantages.* Innovators in the sustainable agriculture/food industry realize that they can gain competitive advantages and new business opportunities by going green.
>
> *Cost reduction.* Use of green practices often enables companies to reduce costs, risks, and liabilities of certain conventional practices, particularly from intensive chemical use.
>
> *Concern about social responsibility.* Companies wish to avoid adverse impacts on health, society, and resources; for them, social and ecological concerns are part of business success.
>
> *Compliance with regulations.* Laws affecting environmental conditions in agriculture have become more strict, inducing change.

are integrated operations; each does production, manufacturing, and/or marketing — and nearly all do business both domestically and internationally, so their influence extends widely.

These cases were chosen to represent a diversity of features, including different sizes and scales of production, within a general sustainable agriculture approach. Despite their differences, each case study shares important common features. The Western region of the U.S. was chosen as a focus of case studies due to limitations in the scope and resources of the project. Many additional cases and other regions could have been included in this report, since there is widespread progress in greening the food system.

Innovators in the greening process are on the cutting edge of contemporary agriculture. These case studies and other similar experiences in this mode share some common motivations and key ingredients (noted in Box 1 and Box 2) that enable progress in greening the food and agriculture sector.

The innovative producers are incorporating basic ecological principles, such as enhancement of diversity (of crops, varieties, soil biota, etc.), recycling and conservation of resources and nutrients, and reduction or elimination of chemical inputs. Most of the innovators in the case studies, and their contracted growers, are also using certified organic methods in at least part of their production, following private or government certification rules. (All U.S. organic crops will be certified under a national organic standard as of late 2002.)

> **BOX 2 — KEY INGREDIENTS OR STRATEGIES OF SUCCESS**
>
> *Leadership* with creativity, vision, commitment, and dedication to principles of sustainability and stewardship that can build team spirit and work hard for change.
>
> *Commitment to sustainability* and the "triple bottom line" - upholding the three interlinked goals of economic profitability, social responsibility, and environmental soundness.
>
> *Innovation and creativity* in ecologically sound and economically viable methods for production, processing, packaging, and marketing, to set new trends and try new approaches.
>
> *Knowledge-intensity in management* of farming and food systems, entailing continual learning, and understanding of complex information beyond chemical inputs.
>
> *Adaptability and diversity,* including adjustment of diverse methods to local ecological conditions, enhancing diversity in varieties and crops, diversifying marketing strategies.
>
> *Gaining value from nature,* taking advantage of natural processes, such as biological functions and organic material, and conserving and recycling resources, to produce high quality.
>
> *Doing more with less,* by enhancing resource efficiency, increasing recycling, and minimizing waste in the food system to increase productivity.
>
> *Forming linkages and partnerships* among companies in the food system, including effective integration between production, processing, and marketing functions, and consumers.

All of these innovators are actively involved in acquiring new information, as well as providing information and services to other growers about sustainable and organic practices. At the same time, they are developing creative and integrated approaches to market their products and meet consumer demands. Some have chosen to scale-up significantly, which creates new challenges and opportunities, whereas others remain relatively small.

PROMISING RESULTS

Green production and marketing strategies often result in multiple benefits and advantages for participating companies in both large and small scales. These methods generally help to mitigate and prevent risks or costs of heavy chemical use, and avoid erosion and degradation of resources. Likewise, they help lessen health risks by reducing chemical exposures and decreasing the chemicals present in the environment. Some innovative techniques make more efficient and effective use of natural processes, and others help to build the natural functions and capacities of organic soils and the durability of productivity.

Sustainable methods also often pay off economically, and in these case studies, they have generally proven to be equally or more productive and profitable than conventional methods over time. In several cases, the annual growth rate of the company's total sales values has exceeded 20 percent in recent years. The overall rapid growth of the national and international organic market — exceeding 20 percent annual growth in the 1990s — is another indicator of economic promise, though growth rates may decrease and stabilize in time as the industry matures.

OBSTACLES AND OPPORTUNITIES

Despite this optimistic outlook, there are major impediments to the continued growth of this green trend in the food and agriculture system. Though the adoption of organic and sustainable practices has expanded rapidly, the total acreage, value, and percentage of sustainably produced food is still very small compared to the values of conventional food and agriculture. Moreover, some of these innovative businesses and particularly small-scale farmers have faced major challenges and downturns from market competition and consolidation of the industry.

Certified organic products, for example, represent only about 2 percent to 3 percent of the total food market in the United States, and generally under 5 percent of market share in western European countries. Why? The report explains that the growth of sustainable and organic systems is thwarted by influential economic, informational, technical, and political factors, as identified below in Box 3.

BOX 3 | **MAIN BARRIERS TO EXPANSION OF SUSTAINABLE FOOD AND AGRICULTURE SYSTEMS**

The main impediments identified in the study are:

Lack of information, research, and institutional support available to producers and other businesses about sustainable practices;

Economic constraints, such as added costs in transitioning from conventional to sustainable practices, coupled with low food prices and market competition that tend to discourage farmers from trying alternatives;

Continued influence of the chemical-intensive model of agriculture;

Inconsistent policy support including contradictory policies that support conventional farming approaches, and lack of policy incentives for sustainable practices.

Equity challenges in the organic industry, including growing market concentration by large-scale corporations, displacement of small-scale businesses, and narrowness in the scope of the organic market, i.e., limited organic food consumption by lower- and middle-income consumers due to higher prices.

Misleading claims about "green" practices by some operations which are actually making minimal modifications.

Still, the outlook is promising for expanding sustainable food and agriculture worldwide. Green approaches offer great opportunities for businesses and for society. However, many stakeholders must take action to overcome obstacles and to accelerate positive trends by increasing adoption of sustainable practices, market opportunities, distribution, relevant research and information access.

Food producers and distributors must realize that the public wants and needs "green" growth or sustainable praces in the food system. Policy-makers also must implement changes, giving greater policy support for sustainable agriculture. In particular, support is needed for sustainable approaches within the U.S. Farm Bill and related legislation. In the United States, decision-makers in the public and private sectors can also learn lessons from Europe about policies that encourage and reward the use of green approaches in farming and food marketing.

Decision-makers are urged to take the necessary actions to expand the sustainable food and agriculture industry, and to overcome the constraints and threats that are being confronted in this sector. Below in Box 4 are five important strategies that must be undertaken by policy-makers, consumers, and producers and other enterprises in the food system.

All actors in the food system can work together on these strategies, and must act now in order to build a truly sustainable food and agriculture industry. The great promise and full potential of this "green" sector can only be realized if barriers and constraints are boldly addressed and overcome.

BOX 4 | RECOMMENDED STRATEGIES:

- *Increase adoption of sustainable farming policies and practices*
- *Build markets and marketing opportunities in the green food system*
- *Increase agroecology research and flow of information about sustainable methods*
- *Prevent the use of environmentally harmful practices*
- *Improve equity and distribution to enable all consumers to have greater access to sustainably produced foods, to protect survival of small farms, and to prevent extreme concentration in the market*

PART I: OVERVIEW OF LESSONS LEARNED

1

INTRODUCTION TO THE FRUITS OF PROGRESS

A green transformation has emerged in the food and agriculture industry in recent years. Growing numbers of farmers and companies in the food and agriculture business are developing environmental stewardship practices and other ecological innovations. Their products and methods are a response to rapidly-rising market demands for food that is produced in environmentally responsible ways and is perceived by consumers to be positive for public health.

These companies use a diversity of green practices, ranging from systematic applications of ecological principles and stewardship practices in all aspects of the business, to minor modifications for experiments with environmental practices. Many of these practices prove to be win-win opportunities—both profitable and ecologically and socially responsible.

Investments are increasing in green approaches to food production, even by venture capitalists and transnational food corporations drawn by potential competitive advantages. This innovative approach to food and agriculture production and marketing can be seen as the state of the art in the twenty-first century.

This remarkable growth of environmental stewardship in the food and agriculture industry is termed "greening the food system" in this report. "Green" refers to not only certified organic practices, but more broadly to a range of methods that are interpreted to be sustainable—win-win approaches that are environmentally sound as well as socially responsible and economically viable *(see Figure 1 and Box 1)*. This report refers not only to innovative food producers, such as growers or farmers, but also food manufacturing and processing companies, marketing businesses, retailers, and consumers, all of whom are linked and integrated in the food system.

Why is there such rapid growth in the use of green practices in food production and marketing? How and why have these green businesses made progress in contrast to the general economic decline of the overall agriculture industry? What are the implications? This report identifies the drivers and motivations behind change, and explains the main factors contributing to growth and marketing success among sustainable food businesses, as well as the ecological and economic results for both private business and public interests. It also identifies

> **BOX 1 | THE MEANING OF SUSTAINABLE AGRICULTURE**
>
> There are many definitions of sustainability in agriculture. But sustainable agriculture is generally defined as a vision of food and fiber production that is economically viable, environmentally sound, and socially responsible and just, as illustrated symbolically below in Figure 1 (Thrupp, 1996, 1998; NRC, 1991; SARE, 2000; UCSAREP, 2000).
>
> We also accept the more specific interpretation in the National Research Council's *1991 Report on Sustainable Agriculture* (NRC, 1991). In this definition, sustainable agriculture is a system of food and fiber that:
>
> a. *Produces food that is safe, wholesome, and nutritious and that promotes human well-being*
> b. *Improves the underlying productivity of natural resources and cropping systems so that farmers can meet increasing levels of demand in concert with population and economic growth*
> c. *Ensures an adequate net farm income to support an acceptable standard of living for farmers while also underwriting the annual investments needed to progressively improve the productivity of soil, water, and other resources;*
> d. *Complies with community norms and meets social expectations.*
>
> In essence, this concept of sustainability is an ideal or a goal, since achieving all of these dimensions can be difficult. Yet, the concept is put into practice through many ways that are explained in this report.
>
> Other terms related to sustainable agriculture approaches include integrated, regenerative, ecological, alternative, biologically based, biodynamic, and of course, organic. These approaches are subsets or particular methods within a broader framework of sustainable agriculture. Those innovators using such green

barriers that hinder progress, and outlines needed actions and opportunities for continued growth of sustainable and organic food systems. By analyzing broad trends in the sector and lessons learned from specific experiences, the report identifies common features and differences among companies involved in the green transformation.

More specifically, the study highlights the characteristics of innovative agricultural and food companies that have achieved significant progress—in financial and ecological terms—by using sustainable practices. Case studies provide one important source of information on these companies. In addition, the report draws upon information beyond these cases, including extensive literature, conferences, and interviews with people who are experienced in this industry. *(See Appendix 1).*

BOX 1 | CONTINUED

approaches are generally working toward goals of sustainability (NRC, 1989; Beus and Dunlap, 1990; NRC, 1991; Thrupp, 1996; USEPA, 1998; SARE 2000, 2001; Hesterman and Thorburn, 1994).

While some growers and consumers believe that organic agriculture is the most "green" or "pure" of these approaches, others point out that certified organic farming is sometimes confined to only input substitution or removal of synthetic chemicals but does not necessarily meet other sustainability criteria. The "Beyond Organic" paradigm represents attempts to incorporate additional stewardship practices along with certified organic, non-chemical methods.

The concept of sustainable business also refers broadly to the intersection of three goals of economic profitability, ecological, and social responsibility for a variety of industries, and has been defined and examined in depth by analysts such as Paul Hawken (1994, 1999), and Arnold and Day (1998).

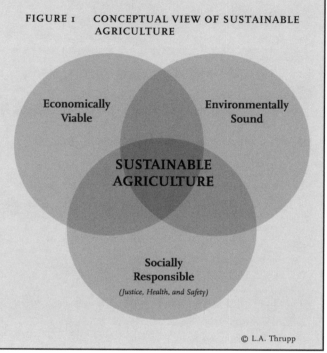

FIGURE 1   CONCEPTUAL VIEW OF SUSTAINABLE AGRICULTURE

© L.A. Thrupp

We interpret the cases selected here as "successful" in a relative sense, recognizing that success evolves over time, and that the companies may have weaknesses as well as strengths. Many additional cases and regions could have been included in this report, since examples of progress toward sustainability exist worldwide (see, e.g., NRC, 1989; UNDP, 1995; Thrupp, 1996; Conway, 1998; USEPA, 1998; SARE, 1998, 2000, 2001; Swezey and Broome, 2000; Corselius, 2001).

The information and guidelines in this report are intended to be useful for decision-makers in the food and agriculture industry, and also for policy-makers and government agencies that influence the food system. *Fruits of Progress* is also relevant for other individuals who are interested in food and farming issues, such as consumers, economic analysts, and others.

2

CHARACTERISTICS AND MOTIVATIONS OF SUSTAINABLE FOOD AND AGRICULTURE ENTERPRISES

AGRICULTURE AT A CROSSROADS: EMERGING 'GREEN' TRENDS IN THE FOOD SYSTEM

Modern agriculture and the food industry in the West has reached a critical crossroads at the dawn of the twenty-first century. While the conventional industrial system of agricultural production has brought remarkable improvements in productivity and a steady supply of inexpensive food to consumers, this system is now under serious stress, and many aspects of it are being questioned. Agricultural producers worldwide are facing unprecedented economic pressures and challenges, in part due to international competition and rising input costs, along with declining productivity and degradation of the natural resources upon which agriculture is based (NRC, 1979; Conway and Pretty, 1991; Conway, 1998; Hawken et al., 1999).

Aggravating these intense problems is a striking pattern of consolidation in the food industry, whereby control of a large and growing portion of the industry is concentrated in the hands of a very few transnational corporations (Heffernan, 1999). At the same time, in certain regions of the United States, over-production dilemmas persist with some crops, exacerbating the depression of producers' prices. Growth rates of the U.S. food and agriculture industry have slowed and nearly stagnated to about 3 percent per year over the last several years (USDA, http://www.ers.usda.gov/).

Together, these pressures and trends are resulting in the bankruptcy and disappearance of thousands of farming businesses, and the collapse of many rural communities (Heffernan, 1999). Meanwhile, in some parts of the world — even among pockets of impoverished populations in the United States — there are serious problems of food insecurity and hunger that debilitate more than a billion people worldwide, creating continual challenges for global agriculture and food distribution systems.

At the same time, after decades of being exempted from environmental regulations, the agriculture industry is coming under increasing pressure by government agencies to mitigate and prevent adverse environmental impacts that have resulted from dependence on chemical-intensive practices, and from continual intensive cultivation of monocultures. Meanwhile, consumers have increasingly demanded foods that are produced with environmentally sound methods and with minimal or no synthetic chemical inputs. These changes are requiring food producers to rethink their conventional modes of using natural resources and chemical inputs.

Amidst this unsettled scenario are signs of hope and inspiration for change. A growing number of farmers and food businesses have recognized that significant changes in industrial agriculture production systems are not only mandated by public agencies and requested by consumers, but can also help sustain productivity and survival of their own farming operations. Recently there have been some significant efforts by producers and others to integrate ecological concerns and green innovations into agricultural operations.

Increasing numbers of producers and food businesses incorporate environmental principles and stewardship practices, and uphold new forms of social responsibility. Throughout the 1990s, there has been a high growth rate of adoption of sustainable practices in the food and agriculture industry. These green innovators are carefully managing resources, exploring new ways of using technologies and chemicals, and developing ways to improve the health of the land and soil.

These innovations have sometimes resulted from attitudinal shifts — from resisting environmental concerns to instead recognizing the value that ecological practices can bring to farming. They are forging a new state of the art by practicing sustainable agriculture, and gaining significant benefits by working with nature—integrating environmental interests into business. (For additional references, see e.g., NRC, 1989; Altieri, 1992; Dunn, 1995; Klinkenborg, 1995; Conway, 1998; Bourne, 1999; Condor, 2000; Ikerd, 2000; SARE, 2000 and 2001; Swezey and Broome, 2000; Greene, 2000a; Corselius et al., 2001; Uphoff, 2001)

What's behind this remarkable green transformation? Although the growth in sustainable practices is a recent phenomenon, organic and sustainable farming approaches are certainly not new. In fact, ecologically based methods have been used in many indigenous and traditional agriculture systems in existence for centuries, starting long before the advent of modern agrochemicals. Some of the principles used in those traditional systems are still relevant and can be adapted in useful ways today (as noted in the Robert Mondavi Winery case, for example).

In contemporary times, ecological agriculture emerged in the United States and Europe as a movement during the 1960s, when some farmers became concerned about agriculture's growing reliance on chemicals and about the ecological and health risks of pesticides and fertilizers. These organic pioneers were often inspired by twentieth-century European farmer/philosophers such as Sir Albert Howard, or by early eco-farming innovators Aldo Leopold, Wendell Berry, and J.I. Rodale (cited in Bourne, 1999, and Guthman, 1999). Later, Rachel Carson's renowned book *"Silent Spring"* appeared to further spread ecological agriculture interests.

These early proponents of ecological farming believed there were disadvantages and costs from the predominant chemical-intensive approaches, even though these production methods often enabled increased yields. These organic pioneers avoided the use of synthetic agrochemicals, and used alternative methods to restore and manage soils to enhance long-term fertility. The early organic advocates were mostly small-scale family farmers who incorporated resource conservation, nutrient recycling, and biological diversity into production; adapted non-uniform techniques to local conditions; and avoided monoculture approaches. They often also promoted broader values of social responsibility and justice, public health protection, and cooperation among people, as well as environmental stewardship and harmony with nature.

During the early years, many saw organic farmers and the health food stores that sold their wares as a small counterculture movement. During the 1970s and 1980s, the movement expanded, and the use of ecological practices spread. In the United States, private and state agencies began acting as third-party certifiers of organic practices, and the federal Organic Food Production Act (OFPA) was passed in 1990 mandating a set of national standards for the organic label. This helped clarify the meaning and credibility of the organic approach, though final standards were not determined until 2001 (for implementation in late 2002).

Under the auspices of OFPA, the National Organic Standards Board (NOSB), an advisory board to USDA, defined organic farming as "an ecological production management system that promotes and enhances biodiversity, biological cycles and soil biological activity. It is based on minimal use of off-farm inputs and on management practices that restore, maintain, and enhance ecological harmony" (quoted by Lipson, 1998).

Certification has often enabled producers to receive a higher (premium) price for organic produce. Since organic production sometimes costs more than conventional measures, mainly due to higher labor costs, this premium price allows the producers to break even, or sometimes to be more profitable than conventional methods. Although 'certified organic' is now legally defined mainly in terms of inputs, many pioneering organic farmers interpret organic farming beyond this strict definition, encompassing ecologically based methods, focus on building soil health and an alternative production system, distinct from the industrial paradigm, as noted in the previous paragraph.

Alongside the development of modern organic farming, "integrated" or alternative "systems" approaches to agriculture have emerged. They have been developed by innovative researchers and farmers trying to decrease reliance on chemical inputs and to find alternative ways of managing pests and soils (Van den Bosch, 1978; Altieri, 1992; Gliessman, 1992; Allen, 1993; Conway, 1998; SARE, 2000; Horne, 2001). These integrated approaches incorporate some of the same concepts and methods as the organic mode, but farmers do not necessarily eliminate synthetic chemicals.

These farmers generally use Integrated Pest Management (IPM), defined here as an ecosystem-based strategy that focuses on long-term prevention of pests or their damage through a combination of techniques such as biological control, habitat manipulation, modification of cultural practices, and use of resistant varieties. Pesticides are used only after monitoring indicates they are needed according to established guidelines. Pest control materials are applied in a manner that minimizes risks to health and the environment. (U.C. IPM program website, citing widely-accepted definition; also see Benbrook, 1996)

Similarly, some farmers have developed alternative methods for integrated soil, crop, and nutrient management, relying more on biologically based methods, and reduced use of chemical fertilizers. In these approaches, the farmers usually manage their systems holistically, attempting to understand the interactions between farm components.

Over time, these integrated and alternative farming methods have come under the umbrella of the term "sustainable agriculture." Other similar terms such as regenerative, ecological, biologically integrated, and even organic, describe diverse interpretations or subsets of this kind of farming, as explained in the Introduction. Early on, growers and scientists developing these methods were a small minority in the food and agriculture community, as was the organic movement. But the methods and ideas they've seeded have sprouted and spread with growing momentum.

REMARKABLE GREEN GROWTH

The development of organic and other sustainable/integrated farming methods boomed in 1989 in the United States, following national news stories about potential health risks from certain chemical pesticides in food, including an emotion-filled story about potential risks to children from a chemical called Alar (daminozide) used in apple production. These media incidents provoked strong and widespread consumer reactions that led to significant growth in organic farming and IPM during the early 1990s. The growth rate continued to accelerate during the rest of the decade. Similar growth also took place in Europe and Japan.

Fruits of Progress does not focus on organic food alone, but it is useful to look at growth in the organic sector, whose specific identity has allowed for some tracking. During the 1990s, the organic industry grew over 20 percent each year in the United States and at similar rates in European and Asian markets. This growth rate far exceeds the track record of the general conventional food market, which has grown only about 4 percent to 5 percent per year during this last decade. The estimated total value of retail sales of organic foods in the United States was $6 billion in 1999 (Dmitri and Richmond, 2000; USDA, 2000; Myers and Rorie, 2000), and $7.8 billion in 2000 (Myers and Rorie, 2000). USDA estimates that a total of about 1.3 million acres were grown under certified organic methods in 49 states in 1997; 63 percent of this total was in cropland, and 37 percent was pastureland. Certified organic acreage grew 111 percent between 1992 and 1997 (Greene, 2000). The growth rate of organic dairy production was especially dramatic: Certified pastureland for production of organic milk grew 469 percent between 1992 and 1997 (Greene, 2000). The estimated total number of organic farmers in the U.S. reached nearly 8,000 in 2000 — an 18 percent increase over the previous year, according to a survey of organic farmers (OFRF, 2000).

The world trade in organic products has also taken off in recent years. In 1997, global retail sales of organic foods reached US$11 billion. Figures for 1998 show that the total was more than US$13 billion. Germany is by far the largest market in Europe (International Trade Center, 1999; IFOAM, 2001), and the percentage of organic acreage (of total agricultural land area) is greatest in Austria (10 percent) and Switzerland (7 percent). Table 1 gives an overview of the estimated size of the major world markets for organic food and beverages in 1997, and expected growth rates of the market.

Although developing countries do not appear in this table, there is also a great increase in production and marketing of organic crops in several of those countries, notably Argentina, Chile, Costa Rica, Ghana, Kenya, Mexico, the Philippines, and South Africa (International Trade Center, 1999; IFOAM, 2001).

TABLE 1. | WORLD MARKETS FOR ORGANIC FOOD AND BEVERAGES, 1997

Market	Approximate Retail sales ($US million)	% of total food sales	Expected growth rate (%) in medium term
United States	4,200	1.2	20–30
Japan	1,000	N/A.	N/A.
Germany	1,800	1.2	5–10
France	720	0.5	20
United Kingdom	450	0.4	25–35
Netherlands	350	1.0	10–15
Switzerland	350	2.0	20–30
Denmark	300	2.5	30–40
Other Europe nations	200 *		
Total in Europe	5,255		

* Included added estimates for Belgium, Finland, Greece, Ireland, Portugal, Spain, Norway
Source: International Trade Center, 1999

This remarkable growth has transformed the organic sector from a small counterculture movement to a multi-million-dollar industry. By the 1990s, organic food had begun to enter the mainstream, even though it still constitutes a minority in the total food sector today (Bourne, 1999; Gilmore, 1999; Condor, 2000; Dmitri and Richmond, 2000; Ikerd, 2000; Ames, 2000a and 2000b). Many newcomers, including conventional food/agriculture producers and distributors, have invested in organic agriculture. In the retail sector, mainstream supermarkets are increasingly adding organic foods while many natural foods stores enjoy considerable financial success. As the market has expanded, some companies have faced constraints, such as difficulties gaining access to conventional markets, and growing competition.

Growers and distributors of organic food have become increasingly sophisticated in their marketing strategies. For example, they produce high-quality foods for gourmet upscale markets and fine restaurants. They serve the growing market of families who want to eat produce grown without synthetic chemicals. Various market options for direct selling of organic produce have evolved in the food system, such as the growth of farmers' markets and community supported agriculture systems (CSAs). CSA is a recent marketing innovation that enables any group of consumers to directly buy produce from a grower through a subscription or membership.

In the late 1990s, the emergence of public concerns about genetic engineering or genetically modified organisms (GMOs) in food amplified consumer demand and interest in organic foods, since organic certification rules do not permit GMOs in farming practices. (This includes most certifier standards currently in use at press time by private and state certifying agents, as well as USDA's federal organic standards, to be implemented in 2002).

The majority of the approximately 8,000 organic farms in the United States remain relatively small-scale and diversified. However, the organic sector has become increasingly consolidated and concentrated, following the pattern of the conventional agriculture/food industry (Buck et al., 1997; Gilmore, 1998; Heffernan, 1999; Hendrickson, 2000; Ikerd, 2000; Lipson, 2000;Uhland, 2000; White, 2000a and 2000b). In the retail and distribution business in the United States, large retail chains such as Whole Foods Market and Wild Oats have captured the largest segment of the natural and organic foods retailing base, and have bought out or shut out many smaller natural and organic food stores. These types of retailers and "supernatural" markets now dominate organic food sales in America; specialty retailers such as Whole Foods Market and Wild Oats account for an estimated 60 percent of total organic food sales (Gilmore, 1998). Likewise, organic farms are subject to the same trend of consolidation, as larger-scale farmers gain a competitive edge, and buy out smaller farms (Buck et al., 1997; Ikerd, 2000; White, 2000a and 2000b). In California alone, only ? percent of the toal organic growers account for over half of the value of organic production. (Klonsky and Tourté, 1993) Venture capitalists have also invested in the organic industry, fueling the tendency toward concentration among well-capitalized firms.

Some small-scale organic pioneers have successfully expanded their operations, taking advantage of the market boom, and evolving into larger companies, including several of those companies included in the *Fruits of Progress* case studies. On the other hand, many smaller-scale organic businesses find it increasingly difficult to compete in face of competition by larger companies. Some have been forced to sell out, unable to survive the changes. These recent trends have therefore changed the structure and face of the organic sector *(see Chapter 4)*.

Alongside farmers working under organic standards are increasing numbers of farmers using a variety of other integrated or ecological farming practices, including IPM and resource conservation methods. The numbers and impacts of these methods are more difficult to ascertain, since they cover such a wide range of practices, and not all are consistently defined and monitored. Nevertheless, it is clear that the adoption of these technologies has greatly increased over the last decade (e.g., NRC, 1989; NRC, 1992; Pretty, 1995; Conway, 1998; SARE, 1998; USEPA, 1998; Corselius, 2000; SARE, 2000; SAREP, 2000; Swezey and Broome, 2000). Some indicators of this trend include:

- Increased growth in sales of biological control agents, compost and other organic soil amendments (Olkowski et al., 1992; Thrupp, 1999).
- A steady increase since 1980 in the publication and distribution of articles, books, and reports on ecological and sustainable agriculture issues (National Agriculture Library, USDA, Mary Gold, personal communication)
- Growth in seminars and training sessions on these issues, and increase in grower attendance at such events.

- Increased acceptance, research, and encouragement in the industry for sustainable and organic practices and by university programs and scientists (interviews with experts, 1999; Thrupp, 1999; SARE, 2000; UCSAREP, 2000).

It has become clear that the greening of the food system involves actions, interactions, and innovations in many functions, and many types of actors working together—from input supply to production and to consumption. (See Figure 2.) Farming, food processing, distribution, sales, and consumer purchasing are all interdependent. Increasingly, these functions and actors have become integrated or merged into single businesses. In other words, companies have developed vertically integrated capacities for production, manufacturing, and marketing.

This type of integration reflects a common pattern in the overall conventional food system. It is nevertheless notable in the green agriculture sector, because it is relatively new among sustainable and organic companies.

In spite of the remarkable boom in green approaches, however, it should be noted that the organic food market is still very small compared to the overall conventional food system. An estimated 3 percent of the total food market in the United States is organically grown produce; and only a small minority of food companies have adopted organic and/or other sustainable approaches. Nevertheless, the growth trend is very strong, and economic analysts predict even higher future growth rates of organic markets and environmental stewardship methods in this sector (Gilmore, 1999; Dmitri and Richmond, 2000; Myers and Rorie, 2000).

FIGURE 2. | **FOOD SYSTEM ROLES**

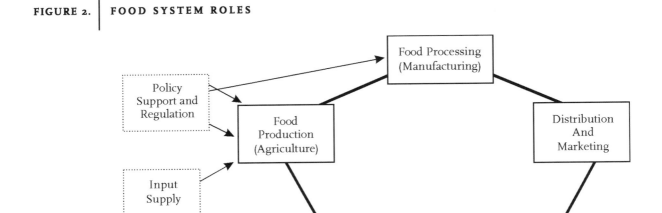

WRI: FRUITS OF PROGRESS

INTRODUCTION TO INNOVATORS

Thousands of food and agriculture companies involved in the green transformation are using a diversity of production and marketing practices that may be sustainable. Diverse approaches and methods are valid for developing sustainability and stewardship in the food industry. Logically, the diverse companies have variations in the speed and extent of adoption of "green" practices. However, some enterprises are only superficially incorporating environmental terms, perhaps to present a public image, while actually continuing to use primarily chemical-intensive methods. Superficial claims of environmental responsibility (a practice sometimes referred to as greenwashing) can have negative implications for companies who are truly involved in sustainable and organic methods, and are potentially harmful to those who are making good-faith genuine efforts, since consumers may challenge the claims.

There are also a growing number of companies seriously committed to a sustainable path that are achieving progress by effectively applying ecologically and socially responsible practices throughout their operations. Because these kinds of initiatives offer useful lessons and insights, twelve established innovators involved in greening the food system were selected as case studies for this report. These innovative operations provide a useful source of information about common elements and differences. (See Table 1 and Case Study Profiles in Part 2 for further information.)

The companies selected as case studies for *Fruits of Progress* share these features:

a. They are upholding principles of sustainability and environmental/social responsibility though there are significant variations in methods and size, and they might not use the term 'sustainable'.

b. All of the cases have vertically integrated operations — meaning growing, manufacturing, and marketing foods — and in a few instances, also create their own materials supplies (such as compost). This means that the companies, and the growers or partners they work with, have adopted and implemented sustainable and ecological practices throughout their operations.

c. They generally have been successful in developing sustainable food production and marketing innovations, where 'success' refers to the simultaneous achievement of economic, environmental, and social aims.

d. They are based in the western region of the United States, primarily California, giving them a common geographical context where there is substantial growth of green innovations in agriculture. Yet, all of these companies have marketing activities far beyond the West: they distribute and buy produce throughout the United States and have international import and export operations. (*See Table 2 for list of companies.*)

Several of the cases are in fruit and vegetable commodities, which illustrate notable progress and are among the highest-value products. Winegrape companies are included in four cases partly because of their active involvement in implementing and encouraging sustainable practices. Examples of grain and row crops are also included to show how this approach works in a diversity of systems.

TABLE 2. NAMES AND FEATURES OF THE CASE STUDIES

Companies/location	Main Products	General Size indicators	Other Features
Del Cabo Farms (*Cabo, Mexico, and Pescadero, CA*)	Tomatoes, Diverse Herbs, and other vegetables	Cooperative with over 250 small-scale farmers in Mexico, & 150-acre farm in CA	All unique organic products, mostly produced in Cabo by small farmers; sold in US
Durst Growers (*Yolo County, CA*)	Various vegetables and melons and alfalfa hay	550-acre farm; contracts with several other growers	All organic products, mostly marketed under Hungry Hollow brand
Fetzer Vineyards (*Mendocino County, CA*)	Winegrapes/wine	3 million cases of wine/year; 80,000 organic cases; over 700 acres organic grapes; contracts with about 300 growers	Fetzer's own vineyards are organic; majority of Fetzer wine is still conventional; ecological practices in winery
Frogs Leap (*Napa Valley, CA*)	Winegrapes/wine	100 acres of grapes, contracts with several growers (100 more acres); 75% certified organic; 50,000 cases of premium wine	Known for producing premium wine from organically grown grapes
Full Belly Farm (*Yolo County, CA*)	80 diverse organic vegetables, fruits, nuts, flowers; and wool/sheep	170 organic acres; 500 members in Community Supported Agriculture (CSA) marketing project	Unique use of biodiversity, including heirloom varieties; strong CSA - direct marketing
Lagier Ranches (*San Joaquin, CA*)	Various kinds of berries, cherries, almonds, winegrapes, citrus; processed fruit spread & desserts	200 acres farmed organically; growing business in processed products.	Creative/diversified marketing and added value processing; recent convert to organic
Lodi Woodbridge Winegrape Commission (*Central Valley, CA*)	Winegrapes/wine	Biologically Integrated Farming collaborative program reaches about 500 growers (whose land covers 40,000+ acres together)	Biologically integrated farming systems approach, such as IPM, reduced inputs, soil management, etc.
Natural Selection Foods (*Carmel & Salinas areas, CA*)	Variety of vegetables, featuring mixed salad lettuces/greens	7000 acres of organic land (with Tanimura & Antle); 2 large processing plants; rapid growth	Majority of produce is organic; sell some of produce is sold as conventional
Lundberg Family Farms (*Richvale, CA*)	12 varieties of rice and numerous rice products (& barley for rotations)	3200 acres of rice, large processing capacity; contract with 25 rice growers (mostly organic)	About one-half of land is organic; other half is "Nutra-farmed" - integrated practices
Robert Mondavi (*Napa Valley, Central Coast, North Coast, CA*)	Winegrapes/wine	7 million cases of wine per year Over 5000 acres of vineyards; contracts with dozens of growers in CA and in other countries	Integrated Pest/Crop management, "natural" winegrowing methods
Sherman Thomas Ranch (*Madera County, CA*)	Almonds, Pistachios, Prunes, Raisin Grapes	700 acres, 75% certified organic, with large dehydrator for drying prunes and rains; contracts with several growers	Recent conversion to biolo-gically integrated and organic practices; sells conventional products at retail store also
Small Planet Foods (*various locations in CA and WA*)	Over 200 processed products, with a variety of processed vegetables, fruits and other foods	$90 million total sales estimated in 2001; growth rate of 20% during late 1990s, contract with over 50 growers & own small farm in WA	One of largest business of organic processed foods; Acquired by General Mills in 2000.

Although these cases have common elements, they also have distinct differences — in scale, types of crops, and production and marketing methods. Cases with a diversity of characteristics were deliberately chosen to show the range of ways that sustainable agriculture is being pursued. Several of the companies are entirely or largely certified organic, while others are not following strictly organic methods, but are using broader environmental stewardship practices that they may describe as 'natural' or 'biologically integrated.' Yet, despite these differences, these cases share some important common features in terms of motivations, factors contributing to progress, and barriers or constraints.

All of these case studies were intended to provide interesting illustrations of leaders in this process, but the group of cases here is not inclusive or representative of all food and agriculture companies on this path. Many additional examples could have been included and would be valuable to profile in future studies, since increasing numbers of growers and food companies are also leaders in green approaches. At the same time, the report is not suggesting that all green ventures are successful, since there are varying results.

WHY ARE AGRICULTURE AND FOOD BUSINESSES ADOPTING SUSTAINABLE PRACTICES?

Why do food and agriculture companies adopt sustainable practices? Why are they turning away from conventional forms of agriculture and toward green innovations? What forces are driving these environmental and green business changes? The transformation is influenced by a unique confluence of causes that include market demand trends, recognition of stewardship values and past mistakes, and realizing new business opportunities that enable competitive advantages.

More specifically, the following factors—six "Cs"— are common primary motivations for adopting sustainable methods. (These factors are similar to those noted in other references about motivations for using sustainable approach, e.g., USEPA, 1998; Walz, 1999.)

COMPELLING CAUSES BEHIND CHANGE

Caring for the Land
Consumer Demand
Competitive Advantages
Cost Reduction
Concern about Social Responsibility
Compliance with Regulations

---

1. **Caring for the Land.** Pioneers in sustainable and organic farming have deep concerns about land stewardship and environmental responsibility. They strive to maintain, conserve, and protect the health of soil and other natural resources, and care about the long-term maintenance of land. They recognize that their own productivity and economic survival depend on stewardship and the health of resources and ecosystems. This attitude about the land and resource base is reflected in their farming methods. Beyond the practical level, these innovators also generally view land stewardship as a philosophy or ethical worldview, reflecting a respect and appreciation of nature in the farming system. Most have an underlying awareness that many conventional agriculture practices

have led to high environmental and economic costs — which they have tried to avoid or curtail.

2. **Consumer Demand.** There has been a rapid rise in consumer demand for reducing synthetic chemicals in foods and for environmental responsibility in agriculture. This motivates businesses to adopt green practices. Public opinion about food quality has proven to be a powerful shaper of business practices, far more influential on markets and on farm practices than in previous times. Consumers continually gain more knowledge about the effects of food production methods and chemical inputs that influence ecological and health conditions (Rodale Press, 1994; Myers and Rorie, 2000; Hartman Group, 1996, 1999, 2000; Jolly, 1991; Onstad, 1999; Dmitri and Richmond, 2000; case study interviews). Public agencies, consumer groups, environmental scientists, and the media are increasingly calling attention to documented problems from certain conventional agriculture practices, such as pollution from agro-chemicals, soil erosion, and pest resistance to pesticides (NRC, 1989; Conway and Pretty, 1991).

At the same time, demand has increased for environmentally sound farming practices as consumers become more and more aware of the benefits of such alternatives. Many surveys show that consumers have a growing interest in food safety, freshness, nutritional value, health aspects, and ecological aspects of food and farming. Many consumers have shown a willingness to pay more for food grown and manufactured using environmentally sound practices (Jolly, 1991; Hartman Group, 1999, 2000; Myers and Rorie, 2000). The businesses highlighted in this report, like other green companies cited in recent studies, realize that consumers' environmental and health interests are opening up market opportunities. These companies also engage in educational and marketing activities to inform consumers about sustainable or organic production, reinforcing and stimulating public demand. Mainstream enterprises that are defensive of conventional farming practices, tend to perceive threats from the health and environmental interests of consumers and public agencies. However, innovative businesses such as those profiled here tend to view green consumer consciousness as a positive force that catalyzes growth of alternative sustainable approaches.

The transformation is influenced by a unique confluence of causes that include market demand trends, recognition of stewardship values and past mistakes, and realizing new business opportunities that enable competitive advantages.

---

3. **Competitive Advantages and Creating Opportunities.** The innovators in this context generally have compelling economic motives for using environmentally sound practices (Arnold et al., 1998). They are motivated to adopt innovations that help them to survive and thrive in the market, to gain competitive

advantages, and to generate revenues to improve the bottom line. Through experience, they see strong and growing market opportunities in natural and organic foods, and realize that revenues can be increased by improving management of natural resources and reducing chemical dependency. Moreover, the growing market in certified organic produce has been lucrative, in part due to price premiums and the emergence of special niche opportunities for high-value gourmet natural foods.

According to a recent USDA analysis, "U.S. producers are turning to organic farming systems as a potential way to lower input costs, decrease reliance on non-renewable resources, capture high-value markets, and premium prices, and boost farm income." (Greene, USDA, 2000, p. 9.) Eco-labeling schemes for green practices also have been developed for alternative production practices that are not organic, increasing competitive market opportunities in some locations.

4. **Cost Reduction.** Companies are motivated by the desire to reduce the costs, risks, and liabilities of certain conventional farming practices, particularly from use of chemicals. Conventional methods, such as intensive pesticide applications, often have high costs, including direct costs of frequent use, and environmental costs that can contribute to production losses (e.g., from pest resistance or destruction of beneficial insects). Some farmers also have been driven to change after experiencing high costs and liabilities from using chemicals that result in pollution. By adopting ecologically based and integrated methods of pest and soil management, growers can reduce the input costs of synthetic pesticides, chemical fertilizers, and their associated costs. Those who significantly reduce pesticide inputs and/or adopt alternatives can avoid liability concerns or criticism by neighboring communities, consumers, and the media.

Adopting these changes is not always easy or cheap. There are often transition costs, such as new management skills and increased labor costs. However, the case studies show that transition costs can be offset over time, and that cost advantages can be significant inducements to use green practices. (See NAS, 1989; Batie and Taylor, 1992; Onstad, 1999; Klonsky and Livingston, 1994; and Corselius et al., 2001 and chapter 2 of this report for more on economics of sustainable production.)

5. **Concern about Social Responsibility.** Several case-study innovators have adopted changes due to genuine concern about protecting or improving social well-being. In particular, they express ethical responsibilities to communities, their children, and/or the wider ecosystems. Leaders or spokespeople say they believe that their company can and should help improve lives and contribute to protection of health and the environment for current and future generations. For example, this social rationale is stated as a motivating force behind Fetzer's commitment to organic and ecological practices. Similar concerns about social and health issues were expressed as important interests for Del Cabo, Full Belly, Lundberg Family Farms, Frog's Leap, and Robert Mondavi as well.

6. **Compliance with Policies and Regulations.** Public pressure from both regulatory agencies and communities have also driven change in some food and agriculture businesses. Prior to 1980, agriculture was not strongly affected by policies and laws concerning environmental impacts. For decades, farmers in many regions have received subsidies for chemicals and support for certain land-use practices, such as continued monoculture planting without crop rotation, that often prove unsustainable in the long run (Young, 1989; Youngberg et al., 1990; Schaller, 1993). However, as public concern has grown regarding adverse environmental and health impacts and risks, federal and state government agencies have increased regulations affecting farm practices and restrictions on pesticide use and pollution. Some agencies have established incentive programs and grants for businesses to encourage voluntary improvements in environmental stewardship. (See Box 2.)

In response to regulatory pressures, much of conventional agribusiness has opposed environmental laws and remained strong defenders of the status quo. In contrast, green food and agriculture companies, such as those profiled in this report, recognize that environmental issues need to be taken seriously for their own economic interests as well as for the public good. These companies have taken proactive approaches to regulations and government policies, by testing stewardship practices, researching alternatives, and using preventive measures before the laws are implemented.

In European countries, government policies generally have had even more influence in promoting change. Several European governments have stronger policies than the United States in terms of both restrictions and incentives affecting agricultural practices. For example, several countries in Europe have established strict restrictions on certain pesticides, and also have generous subsidy programs to support organic and sustainable agriculture efforts.

In sum, several compelling forces and trends have combined to stimulate notable increases in investments and reforms to develop sustainable and organic agriculture. As a result, the green approach is increasingly perceived as a strategically sensible path for many businesses in the food and agriculture industry. The high growth rate is expected to continue in the future, according to food industry experts and economic analysts (Gilmore, 1999; Dmitri and Richmond, USDA, 2000; Myers and Rorie, 2000; Scowcroft, Interview, 2000; Ames, 2000a, and 2000b).

BOX 2   INFLUENCES OF POLICIES AND POLITICS

Policies have been established at the national, state, and international levels that contribute to progress in sustainable agriculture. They include both regulations (sticks) and incentives (carrots):

National Policies and Laws Shaping Sustainable Agriculture

In the United States, a few significant national policies have been established in support of both organic and sustainable agriculture. These policies are administered by a variety of agencies, such as the Environmental Protection Agency (EPA), Food and Drug Administration (FDA), United States Department of Agriculture (USDA), and others. For example, one of the first policy efforts to support sustainable agriculture emerged in 1985, under a few limited provisions of the 1985 Farm Bill. This bill established a Conservation Reserve Program to promote soil conservation, and established a government program known as the Low Input Sustainable Agriculture program (LISA). Although these programs lacked funds, the sustainability issue became a central concern in 1990 Farm Bill discussions (Youngberg et al., 1993). The 1990 Farm Bill included several provisions dealing with natural resources, environmental conservation, and sustainable agriculture (Youngberg et al., 1993). The publication of *Alternative Agriculture*, a major study by the National Research Council of the National Academy of Sciences (NRC, 1989) documented the advantages of sustainable agriculture, based on scientific assessments.

The 1996 Farm Bill contained more provisions to support voluntary conservation measures by farmers and ranchers, including the Environmental Quality Incentives Program (EQIP), administered by the Natural Resources Conservation Service of USDA (NRCS, http://www.nrcs.usda.gov/, 1999). EQIP has provided $200 million each year to farmers and ranchers involved in conservation efforts in the United States, through contracts and grants (Batie, 2000). The 1996 Farm Bill also included a program and funding for protection of farmland and for conservation of ecosystem services (as in wetlands) that might be threatened by agricultural expansion. The Sustainable Agriculture Research and Education Program (SARE) was also strengthened in USDA during the 1990s. Although SARE's budget has been very small compared to other USDA programs, SARE provides public information on sustainable practices, and grants for participatory research and outreach activities to develop sustainable agriculture efforts nationwide (SARE, 1998, 2000).

In 1990, the federal Organic Foods Production Act (OFPA) was passed. This created a national framework for organic production, but many states were already operating under their own standards and laws (Klonsky and Tourte, l994). In 1992, The Secretary of Agriculture appointed the National Organic Standards Board (NOSB), consisting of multiple stakeholders who made recommendations for the national organic law. Following a long and arduous multi-sectoral discussion process, and

BOX 2 CONTINUED

production of drafts that were publicly reviewed and then revised, the USDA released its final National Organic Standards Law in March 2001 for implementation in late 2002, defining uniform rules for foods bearing the organic label.

In recent years, Congress has passed an increasing number of environmental regulations that affect agriculture, particularly restrictions and laws controlling pesticide use, water pollution, and risks from toxic products. The Food Quality Protection Act (FQPA) and the Clean Water Act (CWA) are two significant laws that aim to prevent health hazards, prevent pollution, and restrict the use of certain chemicals. At the same time, EPA, USDA, and other agencies have recently established competitive grants such as the Pesticide Environmental Stewardship program that provides funds to promote pesticide use reduction.

Influential State Regulations and Policies

In California, an important early policy in this field was the California Organic Food Act of 1979 (COFA), which established regulations and formal recognition of organic production in the state. Also, during the 1970s and 1980s, in reaction to growing concerns about agro-chemicals, several new bills regulating pesticides were passed in the state legislature. In 1990, California established one of the most comprehensive and strict pesticide reporting requirements in the United States, obligating farmers to report to the government the quantity and kinds of pesticides used each year. In 1992, the state instituted a mill tax on annual pesticide sales of over $1 billion. This tax currently raises about $25 million each year to help fund the Department of Pesticide Regulation.

Another policy supporting sustainable agriculture in California was the creation of Bill SB872 in 1986 establishing the University of California Sustainable Agriculture Research and Education Program (UCSAREP). This program includes scientific research projects and participatory outreach activities, in collaboration with farmers, on biologically based and ecological farming practices. Other state universities and institutions have created similar programs, such as the Leopold Center in Iowa, and the Kerr Center for Sustainable Agriculture in Kentucky. These kinds of university programs and state agencies, such as California's Department of Pesticide Regulation, have established grant programs to support IPM and pollution reduction. These kinds of changes have also helped motivate some growers to adopt environmental practices.

3

MAIN INGREDIENTS AND STRATEGIES FOR GREENING THE FOOD SYSTEM

MAIN INGREDIENTS OF PROGRESS

What are the primary strategies and practices being used by "green" innovators to apply environmental stewardship and uphold sustainability in the food and agriculture industry? What accounts for progress in sustainable food production and marketing? The case study innovators are using a diversity of approaches. At the same time, however, there are a few main ingredients and strategies that are common to the case studies and provide useful lessons. In general, these approaches improve both environmental and economic performance, serving the companies' self-interest while contributing to broader social benefits. Similar strategies have also been identified in studies of other green business sectors, such as manufacturing (e.g. Hawken, 1994; Johnson, 1996; Arnold and Day, 1998; Hawken et al., 1999), but these qualities have seldom been explicitly recognized in previous analyses of the sustainable food and agriculture industry.

EIGHT COMMON INGREDIENTS OF PROGRESS:

1. Creative Leadership
2. Commitment to Sustainability ("Triple Bottom Line")
3. Knowledge-Intensive Information Management
4. Gaining Value from Nature
5. Adaptability with Diversity
6. Innovation in Production and Marketing
7. Doing More with Less
8. Linkages in the Food System

1. Creative Leadership. "Green" enterprises in the food and agriculture industry tend to have unique and effective leaders with creativity, vision, and dedication to sustainability or stewardship. Their leadership role is critical to enable each enterprise to progress and flourish.

In each case, the primary leader of the innovative initiative is usually the founder, director, chief executive officer, and/or production manager. In several cases, there are two or more joint leaders, such as family members or married couples, as in Natural Selection Foods, Robert Mondavi, Lundberg Faily Farms, Del Cabo, and Full Belly Farm. These leaders are entrepreneurial in spirit, and generally say that they like to try innovations and to are open to changes. Most of them began as farmers directly involved in crop cultivation, and developed many other skills over time. They are eager to learn new ideas and set new business trends in crop production, food processing and/or marketing. Several of these leaders are also recognized leaders beyond their own companies, involved in the wider industry and/or community, often contributing to broader social interests. Another common aspect of their leadership skills is an ability to build team spirit, collaboration, and loyalty among their staff.

2. Commitment to Sustainability ("Triple bottom line"). Success in green business entails a commitment to the goal of sustainability, meaning interrelated aims of economic viability, environmental soundness, and social responsibility, even though the leaders may use slightly different terms to express this innovative approach.

In other words, most of these companies believe that success in the food and agriculture sector entails more than achieving profitability. Effective strategies in these cases are not only economically profitable, but also environmentally sound and socially responsible. The economic viability or net revenue of the operations is logically a driving force, since companies must stay in business in order to be sustained. In addition, they realize the importance of upholding ecological and social responsibility and the positive influence of each on economic returns.

Companies benefit by addressing public concerns and desires about health and ecological issues; by doing so, they may gain consumer trust and increase customer and community satisfaction. Several leaders, including Paul Dolan at Fetzer, Tim Mondavi at Mondavi Winery, Gene Kahn at Small Planet Foods, Larry Jacobs at Del Cabo, John Williams at Frogs Leap, and Judith Redmond at Fully Belly, also express their dedication to broad social or ethical values beyond the scope of company goals. Paul Dolan and Tim Mondavi also mentioned spiritual elements tied to their interests in pursuing broader social and ecological purposes.

For example, an important part of success is establishing good relations, benefits, and just conditions for their employees, and demonstrating respect for the labor force and wider communities. The case study innovators tend to be attentive to protecting the health of workers by providing health and safety training and by reducing human exposure to chemicals or other hazards. (This kind of attention is often evident in conventional agriculture and food businesses, too, but this report does not include a comparative analysis.)

In the case of Fetzer Vineyards, CEO Paul Dolan explains that the company has an expanded concept of the bottom line, which now consists of three interrelated business goals: economics (i.e., financial responsibility), equity (social consciousness and public health), and ecological soundness (environmental responsibility). Dolan calls this the "E-three," a term that Fetzer staff use to describe their overall mission for sustainable business. Fetzer also measures company results according to these aims, using a variety of indicators. This triple-bottom-line concept could be appropriate for many companies committed to a sustainable or green path.

3. Knowledge-Intensive Information Management. By intensively using and managing information and knowledge, rather than relying on chemical-intensive inputs and other uniform technologies, the companies build intellectual capital to develop farming methods.

Increasing numbers of industries are discovering that current and future progress in business is tied to knowledge and information management, and less dependence on material inputs. Companies find that "a key to resource productivity is making creative use of knowledge to drive resource use down, and the value to a customer up," (Arnold and Day, 1998, p. 10). Information is very important in new management systems partly because it creates intellectual capital—a competitive advantage. When companies invest in the use of information over materials, they also reduce reliance on the resources and inputs (such as agro-chemicals) that can be costly and damaging. "In industry after industry, success comes to companies that have the best information or wield it most effectively," (quoted in Arnold and Day, p. 10).

In food production, this is especially true: Decision-makers in these companies are consistently involved in gathering, learning, and applying new ideas for ecological production, marketing improvements, and tracking consumer demands. Green companies tend to substitute intensive use of information and knowledge management for intensive inputs of chemicals, and to adjust practices to specific local conditions. For example, production managers using a biointensive Integrated Pest Management or an organic approach are involved in careful monitoring, observation, and evaluation of data about existing pests, and pest enemies (i.e., beneficial organisms). They often use decision-making models to determine economic thresholds and the appropriate methods and timing for pest control. This approach contrasts with conventional pest control methods, which generally follow standardized and pre-scheduled prescriptions of chemical applications.

Moreover, company managers stress that continual learning is essential for both production and marketing operations. They say they must be knowledgeable and current on sustainable production methods, scientific discoveries about agroecology, and new crop varieties, as well as innovative marketing and processing methods. They engage in research activities on their own farms as one way to advance their knowledge. They also usually incorporate sophisticated computer systems for all aspects of the business, in contrast to early pioneers in the organic movement who often rejected the use of high-tech methods. For example, at Small Planet Foods, managers maintain a state-of-the-art computerized data system that they believe is key for developing successful organic production and marketing operations.

4. Gaining Value from Nature. By carefully managing and taking advantage of natural resources and ecosystem services, the case study companies generate revenue and high quality food products, while protecting natural systems for sustained use.

Successful farming and food production depend on the condition of natural resources and the environment. Soil, water, plants, microorganisms, and other resources in farming systems are valuable natural assets that form the basis of productivity. When soil nutrients and other resources are degraded or depleted, through erosion, intensive tillage, or repeated chemical inputs over time, these natural assets decline or can be permanently lost. The economic value of ecosystem services and natural resources have been underestimated and neglected. To prevent such losses and to avoid high costs of chemical inputs, sustainably managed farming systems make use of ecosystem services, which are functions performed by natural and biological processes (Arnold and Day, 1998, p. 12). The ecosystem services on farms include water filtration and distribution, nutrient recycling and regeneration, climate and air regulation, pollination, natural/biological pest and disease management, and risk reduction from biodiversity. Scientists and producers have recognized that careful management and enhancement of these ecosystem services contributes to overall productivity. (See case studies and Corselius et al., 2001 for examples.)

All of the innovators and leaders interviewed for this study emphasize product quality as a key value for achieving success. They aim to produce the highest quality product as a main business goal, and they perceive their products to have superior quality and taste. They believe that this high quality is largely the result of using natural and organic crop production methods.

5. **Adaptability with Diversity**. By adapting practices and technologies to local ecological conditions, and applying new diverse ideas and methods over time instead of using standardized prescribed technical/chemical inputs, the companies improve efficiency and productivity.

One of the most important and fundamental principles for choosing technologies and methods in a sustainable farming system is adaptation of inputs and techniques to flexibly adjust crop management methods to local environmental and agronomic conditions. This strategy contrasts greatly with the approach generally used in conventional production, where uniform, standardized, and prophylactic technologies or prescriptions are used for applying chemicals and for management decisions. Instead, green agricultural companies modify and adjust their technologies and methods over time, as appropriate, depending on the resources and surrounding conditions.

At the same time, diversity is a key attribute in these operations. Most of these producers grow diverse crops and plants (often in polycultures) and use a variety of different methods that are, again, adjusted to ecological and economic conditions and aims of each case. Even though the winegrape cases are generally monocultures in terms of species, these innovative winegrape producers are increasing diversity in the vineyards in other ways. For example, most of them use methods to conserve and enhance biodiversity, in terms of diversifying crop species and varieties, enhancing diversity of soil microorganisms (which contributes to soil fertility), and attracting diverse beneficial insects (for example, by planting cover crops or conserving border habitat)—all of which have many advantages in production.

In marketing strategies, both adaptability to changing conditions and diversification are also advantageous. Selling a diversity of products in a variety of market channels is a beneficial strategy used in all the cases. A few managers in these cases have also noted that diversity in human resources—a range of capacities and backgrounds—is also part of sustainability (e.g., Fetzer). In general, this incorporation of many kinds of diversity sharply contrasts with most conventional industrial agribusiness features.

6. Innovation. Another crucial feature is the ability to innovate. Managers and other employees tend to be open to learning and interested in trying new ideas. They make creative changes and encourage the use of new sustainable methods in all operations.

These innovators can often turn challenges into opportunities. By adopting new and non-conventional methods, companies and leaders are often going against the grain and taking risks. Yet their willingness to use creativity and to deviate from the norm gives them a new competitive advantage. For example, Larry Jacobs, CEO of Del Cabo explained that he has been able to "use each challenge or problem that the company has encountered to find a new innovation or opportunity." When the company ran into a transport bottleneck due to slow trucks, they found an effective (and inexpensive) means of air transport. Jacobs also invented a new clever package after encountering problems with bruised unpackaged tomatoes.

Innovations are found in both production and marketing practices. For example, in production, changes from conventional to sustainable or organic agriculture involve the adoption of innovative ecologically based methods for resource use and crop/pest/soil management (*See Chapter 3*). Many innovations are tied to contemporary scientific discoveries in organic farming and agroecology. In addition, some producers, including Fetzer, Frog's Leap and Full Belly, have gained ideas from certain traditional knowledge used by earlier generations before chemical use became predominant in agriculture.

Adopting sustainable methods often entails changes in the basic agriculture paradigm and in research and development methods, departing from conventional models. Research is often undertaken on-farm with active farmer participation, and it includes analysis of ecosystem function and of synergies between soils, pests, and crops. Since the people involved in these new discoveries also pass ideas on to other growers, these innovations are being spread widely in the supply chain.

At the same time, these companies creating new marketing, processing, packaging, and promotion methods. In contrast to early organic pioneers, all of these businesses use modern and complex techniques to deal with market conditions. They develop aggressive promotion and sales strategies and make special efforts to respond effectively to customers.

Examples of these innovations include:

a) Manufacturing and selling new processed foods, including unique frozen goods such as organic blueberries and organic TV dinners

(Small Planet Foods), and dried organic tropical fruits and packaged herbs (Del Cabo), convenient rice-based packaged products such as hot rice cereal and crackers (Lundberg), and fresh organic berry pies by Lagier Farms;

b) Widely marketing new and unusual crop varieties that quickly become popular, such as Del Cabo's distribution of extra small organic cherry tomatoes, Natural Selection's sales of mixed organic salad greens, and Robert Mondavi's production of wines from unusual winegrape varieties;

c) Developing creative or unusual packaging or innovative labels (e.g., Natural Selection, Del Cabo, Small Planet Foods, Frog's Leap, Lundberg, and Robert Mondavi);

d) Creating and/or finding new market niches or specialized crops, such as producing kosher packaged vegetables by Natural Selection; fulfilling a "socially conscious" market niche for tomatoes (empowering low income Mexican farmers) by Del Cabo; creating fine quality premium wine from organically grown grapes by Fetzer and Frog's Leap; and successfully developing community supported agriculture marketing by Fully Belly;

e) Developing innovative outreach and/or education programs to reach consumers, other growers, and/or community groups (e.g., LWWC, Natural Selection, Small Planet Foods, Robert Mondavi, Fetzer, and Frog's Leap).

7. Doing More with Less. By improving resource efficiency, increasing recycling, and minimizing waste, the productivity of resources can be maximized.

These innovative food producers and manufacturers are generally reducing their impact on the environment through conservation and efficient use of resources, minimizing waste, and increasing recycling of materials, while at the same time maintaining or increasing yields and returns. These green enterprises have developed promising ways to improve resource management and efficiency. For example, they have benefited from the use of recycled and organic materials, such as agricultural by-products, for making compost, and use manure and other natural mulch materials.

Improving water resource efficiency or reducing water waste is also important, through waste water treatment, recycling, and other conservation measures. The use of natural processes, such as a biological water treatment system for recycled water developed by Fetzer, reduces energy waste and emissions from conventional treatment plants, while conserving and cleaning water. A few are making significant progress in water conservation. Frog's Leap, for example, farms 90% of their land using no irrigation (i.e., "dry farming"). However, water and energy conservation are still major challenges for many businesses converting to sustainable practices, and are critical issues requiring attention.

8. Linkages in the food system. Establishing good working relationships or links with customers, input providers, growers, manufacturers, storage and transport companies, and other marketing businesses is critical to success in a sustainable operation.

The green innovators have developed new and reliable business partners and networks for marketing their products and ensuring that these products reach locations where demand exists. In other words, they must become well-integrated into green food systems and markets in order to thrive. Without these connections, an agriculture business can fail in this endeavor. Although these linkages are also relevant for conventional industrial farming operations, they seem more critical in contemporary green marketing situations, to ensure the company can reach and maintain appropriate customers, business associates, and markets. Most of the companies featured in the case studies have their own capacities for integration. Yet they still must form consistent and reliable linkages and agreements with a variety of actors, from growers to retail buyers. In many cases, these relationships are formalized through contracts, but in other cases, the companies choose to have non-binding links to allow flexibility *(See Chapter 3)*.

PROMISING RESULTS

The growth of sustainable agriculture production and marketing has many positive effects for both private business interests and for broader social and economic interests. From a business perspective, the companies (including growers) undertaking these changes have realized that sustainable practices make good business sense, according to interviews from the case studies and other experts. Although detailed financial data were rarely made available for this study (due to business confidentiality), decision-makers in the case studies confirmed that economic returns have been high, particularly during the 1990s; and sales and total revenues for each have grown steadily, with annual growth rates ranging from 10 to 25 percent for several of these cases. In the case of Small Planet Foods, for example, growth reached 40 percent to 50 percent during some years, and total sales reached above $90 million at the end of 2001.[1] New market opportunities also continue to expand for these operations.

Sustainable methods pay off in production by lowering the costs of chemical inputs, improving and restoring soil quality and ecosystem functions, preventing costs of soil and resource degradation, and protecting worker health, all of which can increase productivity. These methods also can contribute to sustained production over time, by increasing the long-term fertility of soils and the stability of the farming system. Although these kinds of improvements are hard to measure in economic terms, they were cited as important outcomes in several cases. Other studies (such as NAS, 1989, Corselius et al., 2001, Klonsky, 2000, SARE, 2000, 2001; Batie and Taylor, 1989; Ames, 2000a&b), have also documented positive economic outcomes of many cases of sustainable or ecological agriculture systems. More generally, economic vitality of this sector is also reflected in the remarkably high growth rates of the organic market, exceeding 20 percent per annum during the 1990s, as opposed to the conventional food market whose growth remained limited.

Looking at the broader social perspective, one of the main advantages of green practices is reduction or avoidance of chemicals in the environment and of chemical residues in foods. This reduces pollution, decreases risks to public health and to ecological systems, helps protect the safety of workers and communities. Likewise, sustainable methods can improve off-farm environmental stewardship (such as conservation and water conservation) that influence resources in rural communities. These changes improve relationships between these food companies and the public, which ultimately can also help improve the bottom line.

The food companies (or contracted growers) that have switched from conventional to organic often bear transition costs that are higher than conventional costs during the first 2 to 5 years of the change, as noted earlier. These costs include shifting resources from chemical purchases to attaining new skills and knowledge, new management capacities, changes in equipment, and sometimes, an initial decline in yields. Gaining access to new green markets may also add costs during the transition. After the first few years, however, these types of costs decrease and level off, once new methods have been adapted, and knowledge has increased. Companies have confirmed that the costs of adopting new green practices are soon offset by the benefits of change (NRC, 1989; Organic Times, 1992; Buck et al, 1996; Corselius, 2001).

Another outcome of using sustainable methods, mentioned by interviewees in this study, is the high quality of the products, in terms of taste and freshness. The leaders in the case studies claim that they are producing better quality products by using environmentally sound practices and by minimizing or eliminating synthetic chemical inputs. These qualities are not easy to prove since they are subjective. Nevertheless, according to recent surveys and blind taste tests, many consumers also believe that the quality and taste of organic produce are better than that of conventionally produced goods (Hartman, 2000, Natural Foods Merchandiser recent issues, case study interviews).[2]

Although organic produce sold at retail natural food stores in the 1970s was often characterized by inferior aesthetic qualities, great progress has been made in quality control and production methods since that time. Currently, the aesthetic quality of organic products is often equivalent to conventional produce. At the same time, if some products are cosmetically blemished, there are more opportunities to sell them for use in organic processed food. In fact, companies that make processed products, such as the wineries and other cases profiled in this report, often have fewer challenges converting to organic, compared to those selling fresh produce. This is largely because they have less stringent aesthetic requirements.

In contrast to this successful sustainable agriculture sector, the conventional food and agriculture system continues to experience slow and even stagnant growth. Conventional companies that continue to use chemical-intensive production patterns could possibly have more difficulty competing in contemporary and future market conditions. These types of companies also continue, often unintentionally, to degrade resources or disrupt ecosystem functions that are vital natural assets for their own production. Many of these conventional agricultural producers do not yet realize that there is rapidly-growing consumer interest in buying food that is produced in ecologically sound and socially responsible ways. On the other hand, the companies featured in this report, committed to sustainability, have realized the advantages of working with nature's assets, and generally are gaining long-lasting economic benefits. They are cooperating with public agencies and consumers by responding to the ever-growing public demands for environmental responsibility, which gives the competitive advantages in the long run. Of course, these innovators are not free from risks, barriers, or financial downturns since they inevitably face broader market challenges, like all companies in the food industry; but for the most part, they have made notable progress.

Endnotes

1. Pavich Family Farms is an example of another company that had achieved significant success in organic business, reaching sales of $40 million in 1999. But the company encountered significant financial challenges in late 2000, partly due to extremely fast growth in their business, along with market competition and unexpected crop losses. The company's marketing was taken over by another company in 2001.
2. This finding does not imply that conventional foods are "unhealthy" or non-nutritious, but it reflects opinions of surveyed consumers.

4

EFFECTIVE PRACTICES USED IN SUSTAINABLE FOOD PRODUCTION AND MARKETING

COMMON PRODUCTION METHODS

Although the green innovators use a variety of terms and methods (organic, sustainable, regenerative, biointensive), most of them embrace a set of common principles in agricultural production, as summarized in Box 3. These basic principles include biodiversity enhancement, recycling of resources and nutrients, resource conservation, regeneration of soils, and reducing or eliminating synthetic chemicals (e.g., Altieri, 1987, 1992; NRC, 1989; Beus and Dunlap, 1990; Gliessman, 1992; Thrupp, 1996; Conway, 1998; SARE, 2000; Swezey and Broome, 2000; Corselius, 2001; Horne, 2001).

Guided by these general principles, growers in sustainable agriculture use a diversity of farming practices that are adjusted to local conditions, in contrast to the conventional approach that generally uses prescribed inputs and uniform farming methods. Most of the innovators are also using a "farming systems" paradigm which is based on understanding and enhancing the interactions between soils, plants, organisms, and other dimensions of the farming system. They usually emphasize systematic management of information, such as careful monitoring of insects, fungi, crops, climate and other factors — to enable effective adaptation of farming practices.

More specifically, the main types of sustainable and organic production methods used in the case studies (and by many other similar growers) are biologically based or ecological practices for soil, crop, and pest management, as summarized in Table 3. Some of the innovations in this list, such as planting cover crops or habitat strips, have multiple and synergistic functions, and cannot be easily placed in separate boxes. Cover crops, for example, serve several functions, including preventing soil erosion, retaining moisture, improving soil fertility and organic matter, and attracting and harboring beneficial insects (UNDP, 1992).

Certified organic methods are used in many cases, applied in at least half of their land in each case. As noted earlier, certified organic is legally defined in the United States (by federal standards in full effect as of late 2002), in terms of eliminating the use of synthetic chemicals and adhering to guidelines of allowable organic material inputs. While some organic proponents believe that certified organic is the most 'pure'

> **BOX 3 AGROECOLOGICAL PRINCIPLES**
>
> A farming system based on agroecology incorporates the following principles and practices:
>
> **1. Enhancing diversity.** Biodiversity is conserved and enhanced at several levels, including diversity of crop species and varieties, of soil microorganisms, beneficial insects and fungi, useful habitat or plants, variety of cropping systems, and diversified landscapes. Crop rotation and planting various cover crops also enables diversity enhancement over time.
>
> **2. Recycling of resources and nutrients.** Resources and materials are managed and recycled efficiently, particularly soil, water, energy, residues, animal byproducts, green manure, and other biomass from vegetation and habitat that increase the cycling and availability of nutrients, and contribute to productivity and fertility.
>
> **3. Conservation of natural resources.** Natural Resources, including soil, water, and energy, are conserved and protected in various ways to increase efficiency of resource use and to avoid waste and loss of resources.
>
> **4. Regeneration and enhancement of soil.** The soil's fertility, structure, biological activity and overall "soil health" are enhanced by using cover crops, particularly legumes for nitrogen fixation, incorporating compost or animal and green manure, and other means of enhancing biological functions, organic matter, and organisms in the soil.
>
> **5. Increasing synergies in farming systems.** The components of the farm (such as soils, water, pests, and crops) are managed together in a "systems" approach to increase or maximize the interactions and synergies among the components; examples are building soil health that can help crops increase their disease resistance, intercropping to take advantage of compatible plant associations, and managing habitat, trap crops, or cover crops to attract beneficial insects.
>
> **6. Reduced reliance on (or elimination of) synthetic chemical inputs.** Synthetic chemicals are significantly reduced or eliminated, and instead, organic and/or ecologically-based integrated methods of pest and soil management are used.
>
> (Source: Adapted from Altieri, 1987, 1992; NRC, 1989; UNDP, 1995; Thrupp, 1996, 1998.)

and sustainable of ecological farming approaches, this is not always true. Strict compliance to the legal list of certified organic materials alone does not necessarily make a farm system sustainable in all ways. Nevertheless, those using a holistic organic farming approach are more likely to be sustainable than those who merely substitute inputs because they not only follow the input rules, but also use a variety of biologically based management methods noted in Table 3.

VARIATIONS IN PRODUCTION: SCALE AND DIVERSITY ISSUES

The companies featured in this report vary in their practices and approaches to sustainability. One of the most obvious differences is that

TABLE 3.   COMMON METHODS USED IN SUSTAINABLE AGRICULTURE PRODUCTION

Resources or Factors in Production	Examples of Sustainable Practices	Comments and Illustrations in Case Studies (Profiles in Part 2)
Pests, Diseases, Weeds	Certified Organic methods (with no synthetic chemicals) or Biointensive Integrated Pest Management (IPM) methods,* including use of resistant varieties, mating disruption, pheremones, cover crops, crop rotation, trap crops, planting insectaries, introduction of predators, biopesticides, and systematic monitoring of insects & diseases. Selective use of low-risk pesticides for non-organic IPM methods.	Several cases use IPM methods in part or all of their production, and 10 cases use certified organic methods for pest/disease & weed control in over half of their crop production
Soils	Biologically-based soil management practices, including soil "building" with organic materials and/or cover crops, mulching, composting, green manures, using other organic materials, and reduced reliance on (or eliminate) chemical fertilizers; and soil conservation measures	All innovators put a strong emphasis on building healthy soils through biological and organic methods; 5 of the cases (LWWC, NSF, RM, LF, FV) selectively use chemical fertilizers in part of their non-organic acreage
Water	Water conservation, using drip irrigation, dryland farming, or other means, and waste water recycling	At least 3 cases (FL, FV, FBF) are making significant progress in water conservation and/or waste water recycling; the rest are attempting efforts.
Watersheds	Watershed stewardship, by reforestation, vegetation, buffer strips, pollution prevention, etc.	At least 3 cases are developing specific measures for watershed management (RM,FV,LF)
Biodiversity and Vegetation	Cover crops, intercropping, maintaining habitat strips, etc., to attract beneficial insects, reduce or prevent pests, enhance soil fertility, habitat & biodiversity conservation	Nearly all cases are integrating biodiversity in various levels and to varying extents– through cover crops, crop rotation, or other means
Other resources and practices	Recycling of materials & wastes; Conservation of energy, use of "green" (renewable) energy sources	Fetzer has notable progress in recycling and conservation of energy, materials, green buildings, etc.; FBF, FL and other cases attempt this in varying degrees
Worker issues	Prudent Measures for worker safety, rights, benefits, and protection	All case studies include measures for worker safety and protection
Certified Organic Methods	Elimination of synthetic chemicals and use of only inputs identified in the law, following laws and certifiers; also includes many of the methods noted above	8 of the cases (DC,FBF,FL, LF, NS, SPF, DF,ST) use certified organic methods for more than half of their production; 5 cases (DC, FBF, DG, and SPF, LR) produce and market exclusively organic products

Key: DC= Del Cabo, DG= Durst Growers; FV= Fetzer Vineyards, FL = Frogs Leap, FBF = Full Belly Farm, LR = Lanier Ranch; LF = Lundberg Farms, LWWC = Lodi Woodbridge Winegrape Commission, NSF = Natural Selection Foods, RM = Robert Mondavi, SPF = Small Planet Foods, ST= Sherman Thomas Ranch.
* IPM refers to a combination of biological, cultural, mechanical methods for managing insects, weeds, and diseases, and minimizing use of chemical controls. Bio-intensive IPM uses chemicals only as a last resort.

some companies choose to be entirely (or largely) certified organic, while others are pursuing an integrated approach to pest and crop management, or a mixture of both certified organic and conventional. The choice depends on each company's general philosophies, relative costs, risks and benefits, and scale of production. Most of the case study companies have realized significant economic and ecological advantages from being certified organic. At the same time, others, such as Natural Selection Foods and Lundberg Family Farms, also have a significant portion of production in the conventional market, using both strategies for largely economic reasons and market conditions.

However, the organic approach has not worked for every company. For example, Robert Mondavi attempted to convert all its land to certified organic methods in the early 1990s, but dropped the effort after several months, because costs of organic weed management methods became too high to sustain; they switched to integrated methods that still allowed for selective use of herbicides. (Another large winery in California, E&J Gallo, had a similar experience.) Many growers in the Lodi Woodbridge Winegrape Commission also have not fully adopted organic methods, due to what they consider prohibitive costs, and instead use less restrictive integrated pest management methods.

The producers in the cases (and their contracted growers) also vary in the size of units of crop production. Several of them produce in relatively large land areas, consisting of 500 acres or more, but a few also produce in much smaller units. The smaller-scale operations include Frog's Leap, Full Belly Farm, Lagier Ranches and Del Cabo. Several others, including Natural Selection Foods, Small Planet Foods, Robert Mondavi, and Lundberg Family Farms, have large land areas in production. These experiences show that it is possible to grow crops using certified organic and ecological approaches on large-scale as well as small scale farms. However, the very large-scale operations – even if certified organic — tend to be more difficult to maintain and sustain in a truly ecological approach; they are generally less diversified, tend to be managed with more standardized and less holistic methods, and often are more like conventional monocultural industrial operations in many ways. Smaller farm units tend to be more conducive for more genuine ecological organic production, enable more crop diversity and adaptation of agroecological methods to local conditions in a given unit.

The innovators in these cases also approach biodiversity management in varying ways. For example, several of the organic farmers cultivate highly diverse crops within a given plot/space, and rotate crops each season over time. Grape growers, on the other hand, still usually plant one type of crop as a monoculture continuously in a given plot (since they are perennial crops, they are logically much harder to rotate). However, these growers increasingly recognize the risks of relying on monocultures, such as increased vulnerability to pests and diseases, and are diversifying varieties and clones in production, as well as enhancing biodiversity in vineyards by introducing mixed cover crops and vegetation along the borders that contribute to the diversity of soil organisms.

COMMON MARKETING STRATEGIES FOR GREENING BUSINESS

The green innovators have developed various strategies for processing and marketing their products, to both meet and stimulate consumer demands and to increase sales. On the whole, marketing operations in the green agriculture/food enterprise involve some features and approaches that differ from conventional approaches. *(See Table 4 for overview.)*

The ability to change marketing strategies over time is an important characteristic. These companies have had the flexibility to modify approaches, adapting to changing economic conditions and aims. Most of the organic producers initially began by marketing produce, through niche market strategies, to small-scale organic food retail stores, natural food distributors, and/or local farmers' markets.

As the consumer demand increases for organic and sustainably grown food, the companies have diversified their marketing channels and strategies, reaching more customers. Several managers interviewed stated that they often prefer to sell directly to retailers if possible, rather than food distributors and wholesalers, to get a better price. In the special case of those involved in farmers' markets and CSAs, such as Full Belly Farm and Lagier Ranches, the producers market part of their crops directly to consumers. But a few of these "green" producers also market to large reliable distribution companies, since it can be more efficient.

In recent years, a few of the larger-scale innovators involved in the organic market have increasingly focused their main marketing efforts on major natural food chain retail stores, such as Whole Foods Market and Wild Oats, and on mainstream conventional supermarkets. (This does not necessarily reflect a trend for all sustainable/organic farming businesses nationwide, but tends to be true among larger-scale operations.)

All of the innovators have also expanded the geographic range of their marketing. During the 1970s and early 1980s, many of these companies did most of their sales in local markets in the western United States, often selling in urban areas relatively close to their farms. Over time, nearly all of these operations have broadened their reach to other regions of the United States, and abroad. All of the case study companies (except 2) are currently distributing and selling their products throughout the United States and internationally, primarily to Europe and Japan. Several companies, including Del Cabo, Natural Selection Foods, Mondavi, and Small Planet Foods, are increasingly importing crops from Latin American countries, including Mexico, Costa Rica, and Chile, which enables them to have year-round sales. *(See Box 4.)*

Several of the case study innovators have changed over time from using exclusively organic or exclusively conventional methods to marketing both organic and conventional crops. This includes Fetzer, Small Planet Foods, Natural Selection Foods, and Lundberg Family Farms. Several have also partnered with other very large mainstream conventional food companies, partly to expand their capacity to reach mass markets. Small Planet Foods (previously Cascadian Farm) is an example of an organic

BOX 4   EXPANDING MARKETS IN THE GLOBAL ARENA

Many innovators in this "green" food system are expanding their business into other countries by exporting products, and/or importing crops from growers. Small Planet Foods, Lundberg, Robert Mondavi, Fetzer, and Frog's Leap, for example, are increasing sales in European and/or Japanese markets, finding lucrative opportunities to sell sustainably produced foods. In European countries, consumer demand for organic produce is growing at a faster pace than in the United States, primarily due to public concern about food safety, the environment, and health.

Moreover, European countries have national agriculture policies and financial support for converting to organic farming, and the governments have also supported marketing infrastructure and advisory services for organic agriculture, as mentioned earlier (van der Harst, 1997; ITC, 1999, p. 54; IFOAM, 2001). This support has stimulated the greater production and trade of organic produce.

European countries also have uniform standards for organic production and labeling, creating a consistency that is helpful for marketing. Mainstream supermarkets in Europe appear to be more committed to stocking organic products, opening up promising growth. Wine consumers in Europe also tend to both demand and appreciate organic production methods, compared to U.S. consumers, according to Paul Dolan of Fetzer. Given these conditions, it's not surprising that several of these innovators plan to expand their exports.

At the same time, several of the companies are importing more organic crops from developing countries, where tropical conditions provide a consistent year-round supply to markets and consumers in the North. There is little data on the volume, value, and acreage of organic produce being exported from developing countries. But it is known that these export-import markets are

company that started very small in the 1970s, formed a niche in the processed organic food market, and grew exponentially through partnerships and mergers recently culminating in an acquisition by General Mills.

Although their marketing targets have grown, these innovators seldom rely only on singular mainstream markets. They tend to maintain diversified marketing connections, and still try to occupy special niches when possible and opportune. Del Cabo illustrates the benefits of using a niche marketing strategy. Del Cabo has the advantage of providing unusual specialty organic tomatoes and herbs to retail markets during winter months. The organic tomatoes are produced in Mexico at times of the year when it is more expensive and more difficult to produce tomatoes in California in greenhouses. Del Cabo has also developed a good position in the "socially conscious" market niche, since the company works with and benefits small farmers in Mexico, and this contributes to particularly strong sales in certain cities (such as Berkeley, San Francisco, and Seattle) where customers appreciate the social value. However, in recent years, Del Cabo has diversified its marketing, expanded the product line, and broadened its

> **BOX 4   CONTINUED**
>
> growing at double-digit rates (IFOAM, 2001), as mentioned earlier. Some of the fastest-growing markets in organic products from Latin America are organic coffee, cocoa, bananas, tomatoes, and other vegetables. Organic produce in the developing world formerly had to be certified by a certification agent based in the United States or Europe for successful export, but some developing countries are building local certification capacities. (A small number of operations in coffee and cocoa entail "fair-trade" marketing strategies, meaning that the Latin American farmers receive a larger and more equitable proportion of the earnings than in conventional marketing arrangements.)
>
> This pattern of international trade in the green food business has also changed over time. Some ecologically oriented pioneers in this business have refused to export or import goods to/from foreign markets, due partly to their concern about very high energy costs and unfair competition, and instead have stressed localism, urging consumers to support local production and local farmers. Although some, such as Full Belly Farm, have continued that orientation, many producers and food distributors have become increasingly involved in international markets, in response to growing competition and year-round consumer demands.
>
> At the same time, the expansion of trade liberalization policies, globalization, and recent provisions of the World Trade Organization could continue to fuel international expansion in this business. Yet the trend also presents dilemmas, since studies have shown that such expansion tends to escalate market concentration. It can aggravate the gap between haves and have-nots, and potentially can undercut opportunities for small farmers in developing countries to benefit from green growth.

niche, increasingly selling their fresh organic produce to supermarkets and large natural food chains.

Many of the case study companies have also developed innovative ways to add value to their products through unique processing, packaging, and/or new ways of marketing their products. Examples include Small Planet Foods' frozen foods, Del Cabo's clam-shell packaging, and Natural Selection Foods' pre-washed mixed organic greens packaged in plastic bags, and Full Belly Farm's community supported agriculture marketing system.

Several of the companies also use product labels creatively to promote organic or sustainable approaches. Some use colorful artwork to attract consumers and/or include information on the environmental benefits from sustainable or organic production. Certified organic products include the organic certification seal on the label, serving to validate production methods and assure consumers about product claims. Beyond that, some product labels include additional explanations about the ecological and/or quality aspects of growing practices.

TABLE 4. **GREEN MARKETING STRATEGIES AND LINKAGES IN THE FOOD SYSTEM**

Strategies or Methods	Examples in the Case Studies (12 total)
Markets both certified organic and conventionally grown products	5 cases market both organic and "conventional" products (FV, LF, LWWC, NSF, STR)
Market only (or large marjority of) Certified Organic Products	5 cases (DC, DF, FBR, LR, SPF) are all organic
Process food or beverage products (as part of integrated operation and diversification pproaches)	10 cases (DC, FL, FV, LR, LWWC, LF, NSF, RM, ST, SPF)
Penetrate "Mass" mainstream consumer market (eg, supermarkets)	At least 7 cases market in mass consumer markets (DC, FV, LWWC, LF, NSF, RM, SPF)
Include national and international marketing (including export or import to foreign markets)	All cases except FBF have international sales as well as national sales; 3 cases also buy crops from other countries (DC, RM, SPF)
Use Niche marketing approaches	At least 6 cases (DC, DF, FBF, FL, LR, ST) have some kind of niche marketing strategy
Use some Direct Marketing to retailers or consumers	At least 6 cases include direct marketing (DC, FBF, FL, LR, NSF, ST)
Buy produce from other growers (usually contracted farmers)	All cases
Provide information to growers and/or buyers on sustainable/organic farming methods	All cases
Highlight superior product quality as a key marketing feature	Nearly all cases
Undertake outreach activities with community groups, non-profit organizations, or educational institutions	At least 8 cases (DF, FV, FL, FBF, LR, LWWC, NSF, RM) are involved in such activities

Interviewees in the case studies consistently emphasize the quality of their products as an important part of marketing success. They believe their organic or sustainably produced goods have quality advantages, especially taste, and many of their customers also share these opinions. Although these quality descriptions are subjective, many of these innovators believe fervently that the quality of their products gives them a competitive edge in the marketplace.

Methods of determining prices for organic products differ somewhat from pricing schemes used for conventional foods, partly because these products are sold in specialized markets. The companies logically keep close track of changing market prices over the year, and make adjustments accordingly. They pay close attention to consumer opinions, recognizing that consumers are often willing to pay more for organically produced foods. The rate of the premium for organic produce varies greatly, depending on market conditions. In recent years, organic supplies have exceeded demand for some types of products during peak harvest periods. During these times, some companies sell their organic products in conventional markets, receiving no premium.

Innovators that use integrated sustainable methods, but are not certified organic, sometimes use 'ecolabels' or 'green seals' in some regions of the country, such as Oregon's Food Alliance program or the "Core Values" ecolabel program in the Northeast. This acknowledgment of their 'environmentally friendly' practices allows growers to receive a modest premium. However, cases in this report such as Robert Mondavi Winery and many Lodi-Woodbridge winegrape growers are not currently using an 'eco-label' to distinguish their products in the marketplace. Mondavi production managers say that their 'natural' farming approach contributes to the quality of their premium wines, but they do not note their growing method on the label. Yet, some of these growers, like those in the Lodi-Woodbridge group, are considering the possibility of establishing an eco-label for growers who use biologically integrated farming practices. But some in the industry are cautious about eco-labels, mainly due to concern that labels marked IPM or 'biological' can potentially confuse consumers or might raise concerns about products without labels. Others are concerned that eco-labels could impair the meaning and value of the certified organic label. Nevertheless, there is growing interest among some growers and industry groups, to visibly recognize sustainable practices on the product labels.

MANAGING INFORMATION AND RELATIONS IN THE FOOD SYSTEM

As previously noted, a distinguishing characteristic of sustainable approaches is intensive management of knowledge and information. Managers in this business keep close track of information on all aspects of marketing and production. Although information management is also important in conventional agriculture, it is generally given greater attention in green enterprise.

The innovators we studied collect and utilize information from a variety of sources. In addition to doing on-farm research, they also gain new ideas and information from external sources, ranging from articles in scientific and trade journals, to discussions with other growers. Most of them occasionally hire special consultants or scientific advisors in organic and sustainable methods, though these kinds of independent experts are scarce.

At the same time, all the companies regularly provide information and services about sustainable and organic practices to growers, particularly those with whom they regularly contract, and to their customers. They spread knowledge and education to growers in various ways, including farm visits and personal advising by farm managers, group meetings, field demonstrations, and/or distributing newsletters. The Del Cabo case illustrates how providing information provision and education on organic methods to small-scale low-income growers in Mexico was critical to starting the company. In the Lodi Woodbridge Winegrape Commission case, a comprehensive program has been set up for the diffusion, exchange, and monitoring of information about IPM and other integrated practices among growers. Robert Mondavi, Lundberg Family Farms, Frog's Leap, Small Planet Foods, and Fetzer are among those who provide regular advice and visits to their growers, and/or hold grower seminars a few times each year on recent scientific findings and best practices, including sustainable approaches.

This information extension function is part of a broader set of activities, sometimes called supply chain management, for establishing relations with other growers and customers (USEPA, 1998). Forming linkages and good relations with other actors, including growers, distributors, retailers, transport and storage businesses, input suppliers, community groups and public agencies, and ultimately consumers, is important to building a successful green business. Although each of the case study companies has its own integrated capacities for growing, marketing, and often manufacturing foods, each also works closely with other enterprises. These linkages are also relevant for conventional operations, but they can be even more important in the organic and natural food systems, which entail new markets and innovative approaches.

Most case study innovators depend in part on a consistent and reliable crop supply from other growers to achieve marketing goals and commitments. They cultivate a core group of steady suppliers with whom they have contracts. In other cases, they create more flexible or informal buying agreements rather than using formal contracts with growers. The companies not only provide crop production information, but also advice on quality standards, harvest schedules, financial terms, and in the case of organic growers, certification guidelines. The quality control function is often seen as particularly important, to ensure that the growers meet company expectations. Companies also sometimes provide services to growers, such as transport for harvests, produce containers, machinery, and occasionally loans and credit.

These companies, like conventional companies, also interact with government agencies and/or certification agencies, to address regulatory and policy issues. Like all food and agriculture businesses, they are affected by requirements and laws established by the Food and

Drug Administration, the Environmental Protection Agency, Occupational Safety and Health Administration, and the Marketing Standards Board, as well as state laws and agencies. Green companies tend to develop proactive and preventive approaches to deal with environmental laws and issues. Organic companies also maintain close relations with organic certification organizations. In some cases, such as Small Planet Foods, Robert Mondavi, Full Belly Farm, Lundberg Family Farms, and LWWC, the companies' directors or managers have participated in policy discussions or in committees with government agencies concerning organic standards, pesticide laws, land use, and related issues.

Some companies also collaborate with non-profit organizations, community groups, and/or farmer associations for information exchange, environmental activities, or public relations. For example, Robert Mondavi has been actively involved in the Napa Sustainable Winegrowers Group, consisting of a variety of public and private representatives. Full Belly Farm works closely with the community Alliance of Family Farms on projects that benefit the wider farming community. These kinds of connections to public organizations are useful for agriculture industry efforts to develop stewardship and social responsibility, and the organizations can also be a source of useful information for competitive advantage.

5

CHALLENGES AND ACTIONS TO EXPAND PROGRESS

BARRIERS TO PROGRESS

The story is not all rosy. In spite of the remarkable progress and promising efforts described in previous chapters, there are major barriers to the development of sustainable approaches in the agriculture and food industry. A small minority of the nation's growers has made a serious transition to truly sustainable farming practices. Total acreage, value, and percentage of food produced by sustainable practices are still small compared to the value of conventional food production.

For example, although the organic food market has very high growth rates (averaging about 20% in recent years)—that far exceed the conventional food market's growth — its value is still only about 2 percent to 3 percent of the total food market in the United States. Furthermore, although many growers are adopting reduced-risk and integrated pest management methods that prove to be effective and economical, some studies indicate an increase in total pesticide use in recent years (e.g., GAO, 2001; Liebman, 1997). Many mainstream businesses remain reluctant to change the status quo, and some companies currently using sustainable practices have run into hurdles. Conventional chemical-intensive agriculture still dominates the landscape.

Why, when sustainable agriculture seems so promising, is the transition still limited? This chapter summarizes influential barriers to the adoption and spread of sustainable agriculture, clarifying major economic, political, and technological impediments and challenges, which were identified in the case studies and by other experts interviewed in this field. (*See Appendix 1.*)

1. **Economic Barriers and Risks Perceived by Growers.** In general, decision-makers in the farming business logically perceive economic factors as priorities. They must be concerned about their economic situation to ensure survival of their enterprise. Conventional farmers may be reluctant to adopt ecological innovations because they perceive that the economic risks and uncertainties are too high. In particular, they tend to worry about greater labor costs for non-chemical pest control methods, possible losses from pests, and potential sacrifices in crop quality or yields.

Many growers have recently faced serious economic challenges from international competitors, particularly from foreign growers who are exporting low-cost food products to the US. These kinds of pressures aggravate risks and aversion to change by many conventional enterprises.

Growers also face pressures to fulfill marketing standards, government regulations, and food quality and quantity demands from food buyers and packers. Marketing requirements often emphasize cosmetic attributes of food, obligating growers to follow conventional practices to produce uniform results. These marketing standards often prevent farmers from trying alternative practices that they fear might jeopardize their ability to fulfill requirements (Ikerd, 2000).

In many areas, growers also have strong peer pressure from other growers and neighbors to conform to the conventional status quo. For example, farmers are pressured to eliminate all grasses and vegetation in the farm. If they allow natural vegetation to grow between crop rows, and/or plant cover crops, they may be criticized for being messy or lazy by neighbors, though this criticism is unfounded.

Given these economic and social pressures and constraints, fears of economic risk from change are understandable, especially if growers have little previous experience using ecological methods. However, some fears about farming alternatives are inaccurate or exaggerated, due partly to a lack of information on the practices and actual costs and benefits of sustainable and organic agriculture.

Growers who have already made a transition to sustainable and/or organic production generally report satisfaction with the results, as illustrated by the cases in this report. They sometimes face new economic challenges, shifts in costs, and difficulties in finding reliable markets during the transitional years. These initial transition costs are often overcome after two to four years, as growers gain new skills (case study interviews; NAS 1989; SARE, 2000; Corselius, 2001). Beyond this, however, many organic and sustainable growers still face broader economic pressures from market competition, and depression of food prices (Buck et al., 1997). Some also report occasional gaps in the crop supply from contracted growers to fulfill market demand, leading some companies to expand their own production areas. Although organic growers have generally reaped benefits from growing consumer demand, the organic market demand fluctuates for some products, and the organic market could become saturated, some believe.

There are also economic barriers to the expansion of the organic sector from the consumer perspective. Higher prices for organic foods (or for other food marketed with an eco-label) can be a barrier for many consumers who are unable or unwilling to pay extra premiums for food. Although recent consumer surveys show increasing consumer interest in and willingness to pay for organic foods in the United States (reaching 47 percent of consumers, according to Hartmann Group, 2000), a large portion of the population will nevertheless not buy organic foods because of the expense. Most consumers in the U.S. expect to buy cheap food, even though the low prices do not reflect the actual full production costs, and conventional food prices have been kept artificially low through subsidies. (See number 4 below.) Many consumers are unaware of the impacts of their food spending habits and choices on food production and the environment, although their awareness appears to be growing.

2. **Continuing Chemical Dependence.** The continued dependence of the majority of farmers on chemical-based approaches to pest and soil management, and on crop advisers who promote this approach, is a constraint to the spread of sustainable agriculture (Interviews, 1999). Most conventional farmers have become increasingly dependent on agrochemicals over the last four decades, because these chemicals have worked rapidly to control pests and/or boost yields, and because they have been aggressively and widely marketed by their manufacturers and by many pest control advisers. Though there has been growing public recognition and scientific documentation of unexpected high costs and health risks from use of many types of pesticides (NRC, 2000), the chemical-dependent approach has predominated. A combination of factors therefore makes it difficult for producers to alleviate their dependence on chemicals.

In recent years, the pesticide industry has become increasingly involved in research and development of agricultural biotechnology, including genetically modified organisms (GMOs). Growing sales and applications of certain biotechnology innovations in agriculture have created controversy worldwide, and present dilemmas for sustainable and organic agriculture, which are addressed only briefly here. Biotechnology manufacturers, and some farmers and scientists believe that biotechnology offers significant benefits for agriculture, enabling productivity increases and other improvements. Others, including many consumers and scientists, point out that some GMO technologies have ecological and health risks and potentially adverse impacts for farmers and society (see e.g., www.biotechinfo.net; www.ucsusa.org; www.purefood.org; pewagbiotech.org; www.rafi.org). Moreover, some GMOs, such as herbicide-resistant crops, have been developed to purposefully increase the use of certain proprietary herbicides. Although certain GMOs such as *Bt* corn and *Bt* soybeans can potentially help growers reduce standard pesticide sprays, scientists do not fully understand the ecological impacts these *Bt* varieties and *Bt* pollen in the environment. Growing numbers of scientists, consumers groups, food retail companies, and some government leaders, particularly in Europe, have recognized that there is still limited knowledge of the long-term impacts of GMOs. (See Benbrook, 2000; Ervin and Batie, 2000; ESRC, 1999; and websites noted above).

Some farmers and analysts are also concerned that the proprietary nature of GMOs increases the manufacturer's control of production and decreases options for farmers, exacerbating farmer dependency on uniform technologies. Public agencies in most European countries have been more cautious and concerned about GMOs than their American counterparts. These concerns are reflected in restrictions on the sales of GMOs by both retail food companies and government agencies in Europe. In the United States, the development and use of GMOs continues in spite of controversy. However, USDA's national organic standards, in full effect in late 2002, do not allow the use of GMOs in certified organic foods. This regulatory situation as well as public concern over GMOs can potentially increase growth of organic farming and markets.

3. **Information and Institutional Constraints.** Another constraint is the lack of appropriate information, research, and institutional support, by government agencies and by other organizations, for development of sustainable agriculture. Although growing numbers of organizations are becoming involved in sustainable agriculture including organic farming, and a great deal of relevant literature has been published, there are still significant information gaps and institutional weaknesses (Lipson, 1997; Walz, 1999; Sooby, 2001; interviews with growers and other case study representatives, 1999).

Growers and other experts in the field lament a lack of support, research, and information for sustainable agriculture from universities (interviews in case studies and experts noted in Appendix 1). Although universities throughout the United States and abroad have large agriculture departments, numerous research programs, scientists, and extension agents, most of these resources and activities focus on conventional agriculture, and highly specialized disciplines, rarely using systems approaches.

In addition, agricultural research programs and scientists in most American universities have concentrated mainly on chemical forms of pest control, giving relatively little attention to alternative integrated and biologically based pest management methods (Perkins, 1982; Lipson, 1999; Sooby, 2001). Likewise, university and state extension services have usually given relatively little attention to sustainable and organic approaches in their work with farmers (case study interviews, 1999). The reward system for university scientists and extensionists has tended to discourage the multi-disciplinary and integrated systems research approaches preferred in sustainable agriculture. It has also tended to discourage researchers from working with growers (particularly organic growers) on farms. Usually, innovative growers must therefore find independent advisors or alternative information sources.

In recent years, some universities have made some notable changes by establishing programs dedicated to sustainable agriculture, such as SAREP in the University of California and the Leopold Center in Iowa State University. Non-profit organizations such as Appropriate Technology Transfer for Rural Areas (ATTRA), the Organic Farming Research Foundation, the Rodale Institute, and the Ecological Farming Association.

Together, these kinds of programs have increased research and development activities in this field. Such programs have slowly increased the acceptance of sustainable agriculture in university systems, while helping to provide critical information to farmers. However, these programs and the scientists working in them have experienced funding limitations. Additionally, the programs and agroecological principles are often poorly integrated with mainstream agriculture departments.

Federal government agencies with mandates to work on agriculture and natural resources also exhibit institutional weaknesses in this field. For example, the USDA, as well as many state agriculture and environment departments, are relatively weak in their programs and resources dedicated to sustainable forms of agriculture, according to analysts, scientists, and farmers (Youngberg et. al, 1993, Schaller, 1993, interviewees in Appendix 1). For example, organic agriculture research projects received less than 0.1 percent of the total research funding from USDA over a decade of total research funding (Lipson, 1999). Land grant universities have devoted only 0.02 percent (151 acres) of their total research land area for organic experiments (Sooby, 2001).

Recently, some changes have been made by these agencies. USDA, for example, has recently increased its resources and programs in organic and sustainable agriculture. The Sustainable Agriculture Research and Education (SARE) program of USDA is an example of an effective program that supports sustainable farming systems, including some organic research; and SARE is gaining increasing attention within USDA. Yet these efforts are not fully mainstreamed into the parent institution, and they receive minimal funding relative to other programs.

The inadequate distribution of information about sustainable agriculture has contributed to myths and misunderstandings. For example, largely due to misinformation or ignorance, many people (including farmers) believe that environmental stewardship interests and agriculture production interests are inevitably in conflict or oppose each other, though experience proves otherwise. Conventional agribusinesses often do not have sufficient knowledge about the potential profitability of environmentally sound farming practices. This lack of information can exacerbate tensions or perpetuate myths.

4. **Policy Constraints.** Agricultural policies have also presented roadblocks to sustainable agriculture, historically and currently. Over the past four decades, many federal agriculture policies strongly supported and subsidized conventional chemical-intensive agriculture technologies and practices, and impeded farmers from trying alternatives (Young, 1989; Schaller, 1993; Youngberg et al., 1993). Federal programs established in farm bills over 20 years have favored conventional crop production systems using uniform monocultural chemically dependent practices.

For example, the commodity programs of USDA, which existed for over two decades, provided subsidies only for monocultural production of cotton, corn, and other grains, but not for vegetables and fruits (Young, 1989). The 1985 Farm Bill prohibited farmers from rotating crops if they wished to qualify for funding. This obligation to practice monoculture restricts flexibility, thwarts adaptation to local conditions, increases potential for weeds, pests, and diseases, and has contributed to heavy pesticide use. Federal and state water policies also have contributed to unsustainable water use patterns by providing subsidies or greatly discounted rates for intensive water use in agriculture (Reisner, 1993).

Moreover, America's food policy and commodity programs are largely dominated by measures to keep retail food prices low—sometimes known as the "cheap food" policy. These policies perpetuate the continued perception by farmers that chemical-intensive methods are necessary to meet market demands (Youngberg et al., 1993). In addition, common marketing standards, mentioned above, are set partly by regulatory agencies and partly by food buyers and distributors, and tend to thwart innovation by farmers.

Some agricultural policies and programs have been recently changed or rescinded, after studies proved that high costs, inefficiencies, and/or risks in resource use resulted. But the enforcement of such conventional policies for many years has ingrained certain habits, expectations, and outcomes that have been difficult to modify. Though some state and federal agencies have recently established policies that support sustainable agriculture and resource management programs, as described in Chapter 1, these programs and policies have minimal funding and little influence in relation to the support for mainstream policies.

Although numerous pesticide regulations exist on paper, designed to protect health and the environment and to reduce risks and costs to farmers, policy implementation has been weak and often uncoordinated for many of those laws, according to some policy analysts and policy makers (*see list in Appendix 1*). For example, the Food Quality Protection Act was passed in 1996 with the intention of restricting pesticides that pose significant health risks, but the implementation process has been thwarted by lack of resources, by opposition from special interest groups, and by scientific complexities in the reassessment of the pesticides.

5. **Equity and Consolidation Challenges.** The growth of the sustainable and organic agriculture and food sector is characterized by several inequities or imbalances, in terms of increasing concentration in the organic market, limited geographical distribution and a narrow consumer base for organic produce, and farmworker issues explained below. Inequities also exist in conventional agriculture, but they present special challenges to sustainable agriculture proponents since the concept of sustainability is meant to include equity and justice.

In recent years, organic food production and markets have become consolidated and concentrated among a small number of large companies, as mentioned in Chapter 1 (Buck et al., 1997; Dmitri and Richmond, 2000; Ikerd, 2000; Myers and Rorie, 2000; White, 2000a, and 2000b; Lipson, 2000). The consolidation trend is following the pattern in the conventional agriculture industry (Heffernan, 1999; Hendrickson, 2000), and is therefore perceived by some to be inevitable for organic foods as well. Several of the case studies (e.g., Small Planet Foods and Natural Selection Foods) have grown from small family farms to very large companies, and some have benefited from the mainstreaming of organic foods. The managers in these cases generally view this change as an appropriate and dynamic transformation.

Small Planet Foods has gone to the point of merging with a major transnational food corporation, as noted previously. The directors of Small Planet Foods see this merger as an opportunity to improve efficiency and lower the price of organic production, to get more organic produce into mass supermarkets, and to influence mainstream corporations to embrace organic farming. Some retailers and business analysts also feel that large-scale production and mass marketing is a positive trend, spreading organic and natural food more widely in society, and possibly bringing organic food prices down for consumers.

On the other hand, this trend toward consolidation and take-overs by large corporations creates controversies and adverse impacts among smaller-scale businesses and raises concerns about inequities in the food system. Many small-scale businesses in organic farming, including pioneers, have been unable to compete effectively as large businesses have moved in, and many have been purchased or become bankrupt. Due to ethical, economic, and other reasons, many people generally oppose the increasing buyouts and concentration by large-scale corporations in organic agriculture, believing that small-scale operations are preferable and necessary for sustainability and for fairness in the marketplace. Some are concerned not about the size increase alone, but rather, that corporations will alter organic farming to fit an industrial, standardized, input-intensive model that is neither diverse, integrated, nor genuinely organic (Ikerd, 2000; OFRF SCOAR conference, January 2001, Asilomar, CA). Similarly, there is a concern that current consolidation and industrialization changes in the organic industry are eliminating the philosophy and values behind an alternative ecological agriculture system.

The future of small-scale, truly organic operations may be jeopardized in the United States, given these market forces and industrialization trends (Ikerd, 2000; OFRF SCOAR conference, January 2001). Some smaller growers may be able to remain economically competitive in niche markets if they have unique marketing strategies, work in cooperative groups, or if they gain appropriate protection from government agencies or trade associations.

In addition to market inequities, the organic foods market has a relatively uneven geographical spread. Retail outlets and distribution channels are concentrated mainly in large cities in coastal states in the eastern and western United States, and in a few major cities in other regions. The distribution and consumption of organic foods is much less prevalent (or even non-existent) in rural regions and in smaller cities. Although this geographical distribution has expanded recently as supermarkets begin selling more organic foods, significant gaps remain.

Studies also indicate that the consumer base of organic and ecolabeled food is still relatively narrow. Organic consumers are primarily middle-to-high-income and people who have higher levels of education (*Natural Foods Merchandiser* reports; Hartman, 1997-98). Lower-income groups and ethnic minorities tend to be less likely to purchase organic products, in part because of higher prices, lack of access to marketplaces where organics are sold, and/or lack of education and knowledge about organic food. Consequently, many consumers perceive organic and sustainably produced foods as "yuppie" or exclusively for high- and middle-income people. This reputation could potentially be changed if organic food were made accessible to more people through the expansion of direct farmer-to-consumer marketing channels such as farmers' markets, and through changes in targeted marketing and pricing strategies.

In another example of inequity, farmworkers' rights and labor conditions have sometimes been overlooked in organic and sustainable farming, just as in the conventional agriculture industry, according to experts in this field. Although the companies featured in this report have measures and programs aimed to protect worker rights, health, and safety, as noted in the profiles, these aspects have sometimes been given minimal attention by other green growers and food companies. Continued diligence is needed to protect the rights and health of workers as key aspects of ensuring social responsibility and sustainability.

Overcoming and mitigating these imbalances is both a challenge to and a valuable opportunity for the sustainable/organic agriculture industry. Progressive enterprises are already taking steps to address these issues, showing how the meaning of sustainability can be expanded to include social responsibility.

6. **Misleading Green Claims.** Some companies have used green environmental claims or adopted ecological and sustainable terminology, as a public relations strategy, while actually still using conventional approaches. These enterprises may use this "greenwashing" tactic as an attempt to paint a favorable image to consumers or local communities. However, when these superficial assertions are not accompanied by actual significant changes in practices, they are deceptive and can be challenged by consumers or organizations. These misleading claims can also be detrimental to overall greening trend in the food system, and can impair the integrity of those who are pursuing genuine sustainable approaches in the food system.

REFLECTIONS AND RECOMMENDED ACTIONS TO BUILD FUTURE OPPORTUNITIES

Experiences in sustainable agriculture by a growing number of growers and food businesses reveal promising results. The innovators responsible for these inspiring changes are responding to growing consumer demands for food produced in environmentally sound ways. Companies pursuing a direction of sustainable food production and agriculture systems are likely to continue profiting and flourishing, leading the trend and performing on the cutting edge.

These examples show that it is possible to simultaneously fulfill interests of environmental and social responsibility, as well as economic profitability. Even though the cases have variations in their pace, specific styles, and extent of adopting green sustainable methods, this diversity is to be expected among the innovators. Overall, these companies are forging a new sustainable direction in our food system, and developing a new promising business paradigm which includes values that go beyond profit as the exclusive bottom line.

This transition is global, opening up trade opportunities around the world and contributing to the wider distribution of organic and natural foods to diverse consumers. This story of green transition in the food system shows great prospects for future expansion. Undoubtedly, consumers and public organizations will continue to show interest in ecologically responsible production practices and protection of health, as well as affordable and good quality, safe food. Food retailers are likewise responding to this ever-growing demand. Government agencies are also increasing implementation of environmental regulations. This means that food and agriculture companies would be wise to make appropriate changes now in order to respond positively to these trends in the future.

Yet, major obstacles still must be overcome to achieve a broad transition to sustainable agriculture. The great promise and full potential of this sector can only be realized if the barriers are boldly addressed and overcome, and if new alliances and changes are developed. Both the public and private sectors must take concerted actions, explained below, by increasing: (1) wider adoption of sustainable agriculture innovations and policies, (2) marketing opportunities, (3) information access, (4) prevention of environmentally harmful practices, and (5) wider distribution of sustainably produced foods.

1. **Increase adoption of sustainable farming practices and policies.** This is one of the most important and critical challenges. This requires reducing farmer risks (and risk perceptions) in using alternatives to conventional methods. Recommended responses to this challenge are the following:

    Public sector: Increase incentives through subsidies, grants, and educational programs to support the adoption of conservation measures and other sustainable practices by producers, including support to the proposed conservation legislation in the Farm Bill of 2002. One possibility is to require farmers to use sustainable practices in order to receive farm payments or loans. Increase regulations and law enforcement to prevent and penalize the use of environmentally harmful agricultural practices.

Private sector: Increase investments in testing and adopting sustainable practices, particularly natural resource conservation, pollution prevention, pesticide risk/use reduction.

Consumers: Increase purchase and consumption of foods that are produced sustainably or organically, and urge groceries and markets to sell more of these foods.

2. **Increase stable market opportunities in the green food system**, ensuring that new as well as established companies are assured of markets for sustainably produced goods. Recommendations include:

Public Sector: Increase incentives, grants, or loans to protect and encourage small businesses and new entrepreneurs in the organic and sustainable foods business, and help them gain access to capital for this purpose.

Private Sector: Increase investments in activities and enterprises for marketing, distribution, and sales of sustainably produced foods; and encourage private banks to finance growers' investments in developing sustainable practices, giving attention to meeting financial needs of small businesses who are making a transition. To gain access to green market opportunities, smaller businesses need to use creative ideas, including collaborating with other growers by forming strategic marketing alliances and cooperatives, and direct niche marketing.

Consumers: Advocate for greater choice and access to sustainably produced foods in supermarkets and other stores, and raise market demand by increasing purchases of these foods.

3. **Increase agroecology research and the flow of practical information** for growers, and ensure more widely spread and easier access to information on sustainable practices and agroecology. Recommendations include:

Public sector: Increase investments in agroecology research and information diffusion to growers and food companies, and increase information flow on sustainable agriculture from universities, extension systems, and other institutions.

Private Sector: Invest in on-farm research and documentation of results from sustainable methods, and increase the exchange and sharing of information among farmers and businesses, through farmer-to-farmer discussions and other information media. Experience shows the importance of acquiring and tracking new information continuously.

Consumers: Facilitate and build the exchange of information on the meaning of sustainable agriculture among consumers, and urge public agencies and private companies to be transparent and expose full information about agricultural practices.

4. **Prevent the use of environmentally harmful practices in agriculture**

Public sector: Strengthen and coordinate law enforcement measures to prevent the use of environmentally harmful methods; and cease contradictory policies that subsidize or support unsustainable practices.

Private sector: Increase corporate accountability to ensure sustainability and green practices, recognizing that this can improve the bottom line of business, and can add value for marketing; and ensure legitimacy of environmental claims.

Consumers: Call for corporate responsibility by food producers; buy from those who are responsible.

5. **Improve equity and distribution in the sustainable agriculture/food sector,** to ensure that consumers have greater choice and access to such products, to protect survival of small farms, and prevent extreme market concentration. This requires:

Public Sector: Increase education and marketing programs (e.g., through USDA) to expand distribution and access to sustainably produced foods by all consumers in all income levels, e.g., in schools, hospitals, other institutions and workplaces. Establish supportive policies or subsidies to ensure equitable opportunities for small businesses, and apply strict regulations to prevent oligopoly or monopoly of markets by particular corporations in the food system.

Private Sector: Develop and invest in new market channels, and increase sales of sustainably produced food to consumers at all income levels as well as to public institutions such as schools and hospitals. Develop strategies to protect opportunities and competitiveness of small businesses, such as forming cooperatives and alliances and/or other support systems.

Consumers: Help increase consumer knowledge and consumption of sustainably produced foods by supporting farmers' markets, community supported agriculture (CSAs), school programs, and exchanges of information between consumer groups in different regions.

In sum, policy changes, educational programs, and strong proactive efforts by the private sector, public agencies, and consumers are needed to spread the use of "green" approaches, and to support this transformation to sustainable agriculture. All of the above urgent strategies are summarized in Table 5. All actors in the food system, from businesses to consumers, must understand that sustainable practices have advantages for them individually and for the broader society and economy.

Promising results are on the horizon, particularly if these kinds of actions are undertaken. Implementing such transformations widely throughout the economy poses a difficult challenge for the food and agriculture industry and for public sector agencies. Yet the experiences documented in this report show that people and companies working hard and using innovative approaches can be successful and truly make a difference, achieving extraordinary economic, ecological, and social goals. The expansion of these efforts in sustainable agriculture and "green" business in the food industry can create valuable opportunities in the years to come.

TABLE 5. RECOMMENDED ACTIONS FOR EXPANDING SUSTAINABLE FOOD SYSTEMS

Purpose	Private Sector	Public Agencies	Consumers/Citizens
1. Increase adoption of sustainable farming practices	Increase investments in sustainable practices, particularly resource conservation and pollution prevention	Expand incentives, grants, and education programs to support the adoption of sustainable & organic practices; support the Conservation Security Act of the 2002 Farm Bill	Increase consumption, knowledge, and purchases of food produced with sustainable and organic methods, and request grocers to sell more of this kind of food
2. Build markets and marketing opportunities in "green" food system	Increase marketing, distribution, sales and promotion of sustainably-produced foods; link with the natural food business	Enhance incentives and/or grants programs to protect & encourage small businesses/entrepreneurs in the organic, sustainable & natural food business	Expand market demand by increasing purchases of food grown sustainably; advocate for greater choice and access to this food in grocery stores
3. Expand agroecology research and flow of to information about sustainable methods	Invest in on-farm research and documentation of results on sustainable methods; and increase exchange of information among farmers on the info	Increase public funding to research and information-diffusion on sustainable methods, by major agencies, especially in USDA, EPA, FDA	Facilitate and build the exchange and diffusion of information on the meaning of sustainable and organic agriculture among consumers
4. Prevent environmentally harmful practices	Increase corporate accountability for sustain-ability and green practices, recognizing that this can improve the bottom line	Strengthen law enforce-ment to stop environ-mentally harmful methods	Call for corporate responsibility by food producers; buy from those who are responsible.
5. Improve equity and distribution to enable all consumers to have greater access to sustainably produced food products; and to prevent extreme market concentration	Develop new markets and increase distribution to all consumers, and to public institutions such as schools and hospitals; develop support for small businesses, form alliances.	Through education and subsidy programs (eg., in USDA), expand access to sustainably-produced foods by all consumers in all income levels; apply regulations to prevent market monopoly; build opportunities for small businesses.	Increase purchase of sustainably produced foods, and support farmers markets, school programs, and consumer education about sustainable farming and food systems.

PART II: CASE STUDIES

PROFILES OF INNOVATORS

INTRODUCTION TO THE PROFILES

This section consists of brief profiles from the case studies that were undertaken for this report. These summaries provide highlights of the selected innovators who are involved in greening the food system. As noted in Part 1, these twelve cases have been selected to represent a *diversity* of products, approaches, and scales of integrated agriculture and food operations. In general, they are committed to a sustainable path. They all are achieving progress by effectively applying ecologically and socially responsible and economically viable practices in crop production, processing, and marketing. They are based in the Western region of the United States, primarily in California, as a geographical focus.

These diverse cases were chosen partly to show the ways in which agriculture and food marketing can be sustainable and successful in many crops and contexts. In spite of distinctions, these cases share some important common features that have been described in this report. Although some of the cases may seem to be further "ahead" than others in terms of achieving truly green and sustainable approaches, they are not being evaluated or judged in that way. Rather, variations are logically expected and respected among innovators.

The group of cases profiled here is not intended as a complete or inclusive account of all companies involved in developing sustainable green approaches in the food system. Many additional cases could have been included, since there are numerous others involved in this transformation throughout the United States and the world.[1] However, resource constraints and geographical scope limited the study to these few, to provide illustrations of broader green trends in this sector. We appreciate the collaboration of the following innovators for their participation in this study.

Del Cabo
Durst Growers
Fetzer Vineyards
Frog's Leap
Full Belly Farm
Lagier Ranches
Lodi Woodbridge Winegrape Commission
Lundberg Family Farms
Natural Selection Foods
Robert Mondavi Winery
Sherman Thomas Ranch
Small Planet Foods

Dairy operations have not been included in the examples, due to lack of resources, and more difficulty to make comparisons with crops. Nevertheless, there are also growing numbers of dairy operations that are using "green" and organic practices.

PROFILE

Del Cabo

Jacobs Farm/Del Cabo offers a unique model of cooperative organic farming in one country in partnership with management and marketing operations in another, where most of the product is sold.

The company was founded in 1986 by Larry Jacobs and Sandy Belin of Jacobs Farm, a 150-acre family farm in Pescadero, California. A visit to Mexico inspired Jacobs and Belin to help seed a cooperative of farmers there in San Jose del Cabo. Following Jacobs and Belin's vision, the members began growing organic produce and, through Jacobs Farm's marketing and distribution efforts, exported and sold it in the United States during the winter months where those crops were not available as seasonal produce.

From an initial group of eight farmers, the cooperative grew to more than 141 in 1998 and by 2001 had a total of 250 farms, many smaller than five acres. Growing high-quality crops—primarily tomatoes and basil—using certified organic methods and exporting them to United States consumers hungry for fresh flavor in the winter months proved to be a winning strategy. Del Cabo's progressive social mission and organic appeal also created marketing advantages in niche markets.

The company's social mission of building income opportunities for small farmers in poor regions, in addition to using environmentally sound farming methods, was fundamental to achieving this vision.

"Del Cabo was established with the objective of assisting small farmers to improve their economic well-being by teaching them organic agricultural techniques, how to produce specialty crops, and how to administer an organization that would allow them to take advantage of niche export markets in the winter," Jacobs says.

The success of Del Cabo for growers and management alike required more than good intentions. Because members of the cooperative had little experience with either the types of crops Jacobs wanted or the specific methods required for organic certification, the company has conducted a great deal of on-farm training and education. It has taken time to disseminate information on and bring farmers around to the techniques of soil building, composting, use of cover crops, crop rotation, and judicious use of approved inputs. Continued monoculture of tomato crops even under organic certification led to problems with pests and disease that had to be managed carefully.

At the same time, both the cooperative members and Del Cabo's founders and partners in the United States learned by doing, and by making conscious and creative decisions to "turn obstacles into opportunities," in Jacobs' words. Rapid growth wasn't the problem—at times, in the early years, production exceeded the company's ability to distribute. In the 1990s, the company grew at about 20 percent each year, in keeping with the impressive growth of the still-young natural and organic market.

Challenges lay in managing transport of the product, packaging, saturation of the basil market, increased competition, export regulations, and high local labor costs in the Del Cabo region. By creating systems to share risk and guarantee prices for cooperative members each season, Del Cabo has been able to grow and thrive even through some years of torrential rains or hurricanes. The company has expanded its production of a broad diversity of herbs, which have become an increasingly important part of its current business.

The family farmers who make up the cooperative and grow produce for the company earned an average annual income of more than US$20,000 in 2001, compared to $3,000 at the company's start in 1986 (Runsten, personal communication, 2001). This certainly appears to fulfill the objective of Del Cabo's founders to create income opportunities in a developing region. At the same time, the overall business has prospered economically, and their continued adherence to organic standards has ensured environmental integrity.

Del Cabo's experience offers valuable lessons about the achievement of remarkable accomplishments through innovation, perseverance, and creative ability to create unique market opportunities while holding to fundamental principles. By giving serious attention to social, environmental, and financial concerns, Del Cabo offers a model for sustainable development and alternative agriculture and food production.

Information sources for this case study included interviews with Larry Jacobs, John Graham, David Runsten and other staff and growers in Cabo Mexico, and Diana Friedman, 1989, "The Del Cabo Project," Whole Earth Review, Spring. David Runsten's collaboration is greatly appreciated.

PROFILE

Durst Growers

Durst Organic Growers, located in Yolo County, California, illustrates a well-established diversified organic operation that has been recognized by colleagues for leadership and integrity in this field. On their their fourth-generation 550-acre farm, Jim and Deborah Durst organically grow a variety of crops which they sell as fresh produce. Their certified organic produce includes fresh market tomatoes (on about 70 acres), mixed melons, winter squash, summer squash, asparagus, and about half of their acreage is planted in alfalfa hay that is grown for organic dairy feed.

For several decades, Jim Durst's father and grandfather farmed row crops conventionally and also grazed sheep on this farm. After Jim Durst became involved in the operation, he was interested in reducing or avoiding the use of chemicals, so he started farming organically in the early 1980s, beginning with organic wheat. But he faced significant marketing challenges selling the wheat at a premium, due to low demand for organic grains at the time, so they began to convert some of their acreage to vegetables. The company improved its skills in organic farming, and became successful in producing organic vegetables as well as alfalfa. By 1991, they had fully transformed the farm into a certified organic operation.

Central to the Dursts' farming practices is building soil fertility and balancing insect ecology. Jim Durst says he focuses on improving soil conditions "to create a healthy environment for the organisms that help our crops..." Jim believes in the principle: "feed the soil, and the soil will feed the plants." One of the main practices that they have used since the 1980s is cover crops, which have given them multiple benefits, such as adding "biomass to the soil... improved soil structure, water permeability, and a healthy soil fauna." They also use crop rotation and other methods to avoid soil disturbance and compaction.

The Dursts take steps to create what they call "a healthy work environment" for people who work on the farm. They have several full-time employees, and they also hire more than 80 workers during the harvest periods. To these seasonal employees, they also offer a retirement plan and provide worker training opportunities. The employees are engaged in multiple aspects of the farm work, gaining skills in many jobs. The Dursts believe it is useful for the individual employees and for the entire team to have a better understanding of how the whole farm works in order to improve judgement and versatility.

The Dursts market their vegetables under the brand name Hungry Hollow to wholesale and retail outlets in the San Francisco Bay area and to many cities in the United States and in Canada. Starting about two years ago, they began using an agent called "Organic Harvest Network" which helps them market and promote their fresh produce nationwide to wholesalers and retailers. The Dursts also partner with several other growers in their area whose produce they buy and market under the Hungry Hollow label. The volume of produce that they buy from these growers varies year by year, depending on market conditions and quality of produce. The Dursts want to ensure their buyers the best quality products, and work with their growers to try to achieve this result.

The Dursts also participated for several years in a research project with the Sustainable Farming Systems (SAFS) project at the University of California at Davis. In this project, Jim Durst cooperated closely with scientists who analyzed several kinds of farming systems, and did detailed on-farm measurements of crop yields, soil quality, economic returns, and overall sustainability. The project team of scientists and growers met fairly often to share knowledge about the results, hold field days, and also distribute the findings to the agriculture community in Yolo county and beyond. Durst and others involved in this project feel that this kind of research and information-exchange from the on-farm experiments are very useful for growers, and also helps to spread positive change over time.

The Durst Growers have chosen a path of slow and steady growth in their business. They deliberately have not expanded their own land area, remaining at 550 acres, and they don't intend to buy more land, since they prefer to remain solid and sustainable at the current size. However, they have also focused on improving the quality of their products and increasing the efficiency and extent of their marketing. The Durst Growers will continue to provide exemplary leadership for other growers in the area who are pursuing sustainable and organic methods.

Information sources for this case study included: Interviews with Jim and Deborah Durst; Dr. Stephen Temple, U.C. Davis; Organic Harvest website; and secondary sources, including summary in Western SARE, 2000, "Sustainable Agriculture... Continuing to Grow," Western Region Sustainable Agriculture Research and Education Program, with Sustainable Northwest, Portland, OR, pp. 44-45.

PROFILE

Fetzer Vineyards

Fetzer Vineyards has been recognized as a leader in developing and implementing environmental innovations throughout its large-scale enterprise. Based in Mendocino county, about 80 percent of Fetzer's own vineyards are certified organic. The company encourages similar practices among its 300+ contract growers. In 1993, with its Bonterra brand, Fetzer was the first major American winery to develop a premium wine made from organically grown grapes.

Fetzer's environmental stewardship goes far beyond its vineyards to include notable achievements in ecologically sound building construction, a comprehensive recycling and resource conservation program, and an emphasis on human values in the workplace. Under the leadership of CEO Paul Dolan, the company is committed to three inter-linked goals of economics, ecology, and equity that they call the "triple bottom line." Fetzer's staff has created an "E3 Team" to coordinate and help implement these triple goals that guide their sustainable business practices. Fetzer's mission statement reads: "We are an environmentally and socially conscious grower, producer, and marketer of wines of the highest quality and value. Working together in harmony and with respect for the human spirit, we are committed to sharing information about the enjoyment of wine and food in a lifestyle of moderation and responsibility."

Founded by the Fetzer family in 1968, the winery and vineyards are located in a fertile valley at the base of the coastal range foothills of Mendocino County in northern California. The company was sold to the Kentucky-based Brown Foreman Corporation in 1992. Today, Fetzer farms about 700 acres of land, and contracts with approximately 300 growers. About 80 percent of the company's land is certified organic, with more in transition from conventional to organic. The company employs about 300 people. Fetzer sells about 3 million cases of wine per year in all 50 states and in 25 other countries. Sales of the Bonterra brand constitute about 3 percent of Fetzer's total sales volume and have been growing at a fast rate in recent years.

Fetzer's commitment to organic and environmental practices began in earnest in the mid-1980s, when the company established a five-acre vegetable and herb teaching garden for culinary classes and for catered events and visitors. The garden's manager, Michael Maltas, applied his knowledge of biodynamic and organic farming practices. Positive results in the garden led vineyard managers to begin incorporating organic practices in the vineyards. They continually expanded the vineyard area under organic methods, and convinced by growing successful outcomes, they soon became certified by CCOF. They use diverse cover crops and compost as key practices to build soil health which Fetzer views as a critical basis of a healthy and productive organic system.

At the same time, Fetzer began a dynamic recycling program that has grown over time. Each year, the company recycles at least 70 tons of corrugated cardboard, 75 cubic yards of paper board, 500 tons of glass, 740 gallons of oil, and 392 cubic yards of wooden pallets, according to their web site. In addition, 12 cubic yards of cork and 10,000 tons of grape pomace are composted each year. The company has reduced its discarded materials by 93 percent between 1990 and 1997, eliminating dumping of 1,580 cubic yards of landfill. Fetzer has won Waste Reduction Awards of the Year from the Waste Management Board of California's Environmental Protection Agency. In 1997, Fetzer was given a special Waste Reduction honor, acknowledged as one of the ten best recycling companies in California.

CEO Paul Dolan, inspired by Paul Hawken's *The Ecology of Commerce*, began looking for more ways that the company could address global ecological issues. During the mid 1990s, they initiated construction of ecologically sound winery offices using recycled building materials, enhanced energy and resource conservation, and other "green" building practices. Fetzer's winemaking, storage, bottling, and labeling practices also use ecologically innovative methods. The company also recycles and reuses water from the winery. Water treatment ponds with electric aeration mechanisms, and a unique state-of-the-art biological treatment pond uses natural vegetation, in which cattails form a natural filtration mat in the water. The system reduces energy input for the aeration system while effectively treating water through the natural microbial activity of plant and water bacteria. Also, in late 1999, Fetzer began using "green" energy utility services based on renewable energy sources.

Fetzer also promotes a management style that is relatively non-hierarchical, informal, open, and supportive of staff. Dolan places a premium on good communication and information flow. All employees in the company are encouraged to set their personal job performance goals in relation to the company's "ecological, economic, and equity" mission. Vineyard managers stress that management of information and knowledge is more important for organic production than for conventional farming, particularly for monitoring and evaluating variations in climate conditions, soils, pests and diseases, and for adjusting practices to localized needs.

Fetzer's challenges have included garnering the interest, enthusiasm, and support of those who have not yet embraced environmental and social sustainability as a driving force of business. Although the company's volume of Bonterra wine is growing rapidly, organic wine is still a minor percentage of the overall production. From growers to investors, many people still perceive organic or other ecological practices as high-risk. Dolan says he will continue to improve understanding about organic approaches through education. Fetzer's future goals include further reduction of energy use in transportation and winery operations, expansion of organic practices among their contracted growers, and strengthened staff training, diversity, and community involvement.

Information sources for this case study include: Interviews with Paul Dolan, Tom Piper, Scott Duncan, Mike Johnson, Bill Cascio and other staff from Fetzer; website information; discussions with green business collaborators; Kohn Properties staff.

PROFILE

Frog's Leap

Frog's Leap, located in the heart of the Napa Valley in California, produces premium wines using ecological practices in grapegrowing and winemaking. Although Frog's Leap is much smaller than some other companies in this report, it is notable as a pioneer in producing high quality wines from organically grown grapes, and for its unique niche and sustain- ability in the business. Frog's Leap currently farms about 200 acres of winegrapes and operates a winery in Rutherford.

Frog's Leap winery was started in 1981 by its CEO and winemaker John Williams, who initially purchased grapes from local growers to make wine. In the late 1980s, Frog's Leap acquired 15 acres of land and began growing grapes. Williams began trying ecological practices in this vineyard because it made common sense to him. While experimenting with various methods, he discovered that the organically grown grapes generally produced better flavor and a better quality wine, in his opinion. He continued to develop this organic approach, and his land was certified by CCOF in 1989. Through the 1990s, Frog's Leap expanded the area they own and farm, and continued to refine ecological methods that Williams believes are best for "healthy wine growing" and for making high quality wines.

Today Frog's Leap owns about 100 acres of certified organic land, and they contract with other growers in another 100 acres. About 75 percent of the total land they farm is certified organic, and the remaining acreage is farmed ecologically or in transition, but is not certified. John Williams explains that the goal of the company is "to produce wines that deeply reflect the soils and climate from which they emanate." Frog's Leap also has a special reputation due to their sense of humor and fun spirit, reflected in their clever labels and amusing promotional activities.

Williams emphasizes the importance of healthy soil in his organic system. The company uses a variety of cover crops, which enhance the soil organic matter and microbial activity, improve soil structure and fertility, and can suppress weeds and attract beneficial insects. When leading tours, Williams shares his excitement about the natural qualities in the vineyard by encouraging visitors to "smell the soil "– which generates a complex aroma that is alive with organic life. Another unique feature of Frog's Leap's ecological approach is that a large majority of the land they farm is "dry farmed," meaning that it is rainfed, not irrigated. Frog's Leap also incorporates other basic principles of "recycle, reuse, and renew" in their vineyard operations.

The winery also uses natural principles in winemaking--using natural yeasts, minimizing handling, and avoiding unnecessary filtrations of the wine. Williams has begun experimenting with biodynamic farming principles, mainly in the landscaping and in their mixed vegetable, herb, and fruit garden that is attractively located in front of the winery. (This entails the integration of natural cycles, rhythms and special biological treatments, understanding and managing the farm as a complete living organism.)

Frog's Leap's production has remained steady in recent years, at about 50,000 cases per year, while the marketing range has broadened, and their profitability and product diversity have increased over time. The company has matured and grown internally by increasing infrastructure, work force, and its land ownership. Frog's Leap niche markets their wines mainly in fine restaurants and wine shops in California and in selected cities throughout the United States. The company also exports wine to Europe, Japan, and Canada. Although their exports represent only about 8 percent of total revenues, sales abroad have grown steadily.

Frog's Leap's label does not indicate that the wine is made from organically grown grapes, partly because a small portion of the grapes they process are from the non-certified land of their growers. Williams also believes that using the organic label currently does not generally give their wine a market advantage or a premium in the wine market. Some of their customers appreciate Frogs Leap's ecological orientation, but other customers care most about the fine quality or taste, and pay more for the wine based on that aspect alone. Yet Williams is convinced that the use of organic methods contributes directly to higher quality and flavors of their wine.

Frog's Leap has 30 full time employees and hires additional seasonal workers during harvest. Many of the employees have been with the company for over 10 years, and are committed to its values. Williams has helped to train and mentor several young employees, including Mexican American workers, who have now developed professional expertise in winemaking and other skills.

Williams and other Frog's Leap staff are involved in supporting local environmental and educational activities in the community, such as providing field tours, seminars, and giving talks at conferences. The company's efforts can help increase growers' and the public's understanding of why and how to care for the soil and other resources, to create healthier vineyards and high quality products over the long term.

Information sources for this case study included interviews with John Williams, CEO, and Frank Leeds, farm manager; information from Amigo/Bob Cantisano, organic agriculture consultant; and Klinkenborg, Verlyn, 1995, "A Farming Revolution: Sustainable Agriculture," Natural Geographic, December, pp. 61-88.

PROFILE

Full Belly Farm

Nestled in the Capay Valley of Northern California, Full Belly Farm is a well-established and successful organic farm that is known in the region for its innovative marketing and progressive employee relations, as well as for growing and marketing a very broad diversity of vegetables, fruits, nuts, and flowers all year round. Although this farm is smaller in scale -- 170 acres -- than many of the other cases in this report, it is highly productive and its gross revenues have grown steadily at a rate of 10 to 15 percent per year over the last decade. Full Belly harvests more than 80 kinds of organic crops and also maintains about 100 sheep (and a few other farm animals) that have very important functions in their integrated operation. The farm has a reputation as a mentor and supporter for other small-scale organic farmers.

Full Belly Farm was started in 1989 by four partners/owners, Andrew Brait, Paul Muller, Judith Redmond, and Dru Rivers. The land has been farmed organically since 1984, and is certified by CCOF. In addition to avoiding synthetic chemicals, they use cover crops that fix nitrogen and provide organic matter for the soil, apply compost, and plant habitat areas for beneficial insects. The sheep are a valuable part of the operation, managed in a rotational grazing pattern on the farm. They graze on crop residues and on cover crops, which enables mowing of the vegetation to create useful green manure that is incorporated directly into the soil. The farm also sells the wool and sheepskin.

The owners stress the value of biological diversity in the farm. Growing many crops and varieties help to prevent diseases and pests, and diverse cover crops and surrounding habitat also increase the health of the soil and the system. Having a broad diversity of crops, including heirloom varieties, also appeals to their customers. They also make sundried tomatoes that are sold in the summertime. Farming operations are continued year round, even through the winter, when they continue to produce vegetables such as greens and coles.

The farm has approximately 25 employees, most of whom are retained all year. This kind of year-long employment is unusual in a small farm, since most farms this size only have a very small handful of permanent employees and then hire temporary seasonal employees during harvests. Full Belly Farm also has an apprenticeship program, which helps build knowledge and capacities of young adults. Each year, they usually hire a small group of apprentices, who live on the farm with the owners and their families, and take part in the farm's unique community. Judith Redmond explains proudly that several of these apprentices and other employees have been inspired to continue farming as a career, and some former employees have recently started their own small farms in the area, with mentoring and support from the Full Belly owners.

Full Belly's products are sold mostly in California, but sometimes reach other states through their wholesale distributors. The company has a diversified marketing strategy: Their products are marketed to retail stores and restaurants (accounting for about 33% of their gross revenues), wholesalers (about 20%), farmers' markets in the San Francisco Bay Area (also about 20%), and through an innovative form of marketing called Community Supported Agriculture (or CSA), which they began in 1992.

For their CSA, Full Belly prepares boxes of fresh produce every week that are distributed directly to 'subscribers' who are members, mostly in Davis, Sacramento, and the Bay Area. The program has achieved significant success, and members have increased steadily over time. Now, the Full Belly CSA accounts for more than 20 percent of their farm's total gross revenue, and they have 500 members. During the winter, Full Belly also includes fresh organic oranges in the CSA boxes that they purchase through an agreement with a neighboring farmer, since Full Belly does not grow oranges. Judith Redmond believes that the CSA system has special qualities, such as enabling consumers to directly connect with the farmers, and also because the customers appreciate the local supply of very fresh seasonal food. However, Full Belly also continues to market more than half of its produce in other retail channels, since not everyone has access to or prefers the CSA system.

Full Belly Farm is actively involved in community educational events about sustainable and organic farming through school visits and tours for visitors. The farm also puts on an annual harvest festival for the community and other farmers, and the celebration usually attracts hundreds of people. The farm owners have also been active in political efforts related to developing policies to support sustainable and organic farming.

Information sources include: interview with Judith Redmond; Ecological Farming conference in Asilomar, CA 2000-2002; and other experts in sustainable/organic farming; website; and Community Alliance of Family Farmers.

PROFILE

Lagier Ranches

Lagier Ranches, located in Escalon, California, in San Joaquin County, has family roots going back to the late 1800s in the region. As a fourth generation farmer and entrepreneur, John Lagier has developed a new path in his family history: His recent experience illustrates a successful example of diversification, organic conversion, and innovation in manufacturing products and direct niche marketing.

Lagier's parents and grandparents historically farmed row crops and almond orchards using conventional methods, and also raised mules. John Lagier began farming in 1979, partly by leasing land from other family members, and he began a process of diversification and innovation that has continued up to the present day. He established vineyards and cherry orchards, and then in 1991, began to convert the 200 acres he farmed to organic methods. He adopted these changes partly due to health concerns that he and his wife had about use of chemicals, since both of them had experienced cancers. Given his dedication to change, the conversion was quite rapid, and all of the land he farmed became fully certified organic by 1997. Lagier Ranches currently produces a variety of organic berries, almonds, cherries, winegrapes, and a small area of citrus and exotic crops such as pawpaw. Their organic practices include the use of diverse cover crops, compost, foliar feeding, and minimum or no-tillage throughout the farm.

Lagier Ranches has developed diverse strategies for processing and marketing. The company does in-house manufacturing of several added-value products in a commercial kitchen that Lagier recently established on the farm. They manufacture organic products that are made mostly from their own crops, including fruit spread, almond butter, almond snacks, and pies. They also purchase a few additional ingredients, such as evaporated cane juice and organic wheat, mostly from local businesses. Lagier explains that they originally started this manufacturing mainly because their berries are highly perishable, so they can avoid losses of fresh fruit by processing it. Although their manufactured products currently make up only about 15 percent of their total sales, developing this processing capacity adds value and is beneficial for their business.

The company's fresh produce is marketed to a variety of places, such as retailers (including local fruit stands), and wholesalers, who distribute in California and in other cities on the East Coast and the Midwest. They also retail their produce in several farmers' markets in the San Francisco Bay Area. Most of their manufactured products go to natural food grocery stores such as Whole Foods, and to other U.S. retailers or distributors who appreciate these products; a small amount is sold to a distributor in Japan. The almonds are hulled and shelled through a local processor, and they work with a cooperative of organic almond growers in Turlock for packing and selling them.

For Lagier Ranches, the organic conversion and diversification process has been relatively smooth, even though there were some risks during the transition period due to yield losses. Once the acreage became certified and they began to earn a premium on the products, they have seen considerable economic and environmental advantages compared to the conventional systems. Lagier explains that the paradigm switch to organic required new learning and new information which he has gained largely through other growers who are open to sharing their knowledge.

The biggest production challenge they face is gophers, and they are using a variety of methods, including trapping and owl boxes, to control them. The depressed prices in the overall market situation recently has inevitably affected the operation, but not dramatically. Lagier stresses that organic growers like himself tend to be better off than conventional growers and food companies under current market conditions. The company continues to be committed to this way of farming, and dedicated to refining and growing their innovative marketing approaches in response to market conditions.

Information sources include: Interviews with John Lagier, founder, and Matt Devator, production manager, and Cindy Lashbrook; presentations at the Ecological Farming Conference in January 2002, Asilomar; "Partnerships for Sustaining California Agriculture" conference in Woodland, 2001; and company website.

PROFILE

Lodi Woodbridge Winegrape Commission

The Lodi Woodbridge Winegrape Commission (LWWC) is an association of 650 grape growers in the Sacramento River Delta region of California. In all, the members of LWWC cultivate more than 70,000 acres, making the region a leading producer of winegrapes in California. LWWC is notable for its integrated farming program, in which many of its member growers participate. Through grower-based educational and outreach activities, the program is successfully implementing innovative pest management methods, reducing agro-chemical inputs by many growers in the region, and carrying out on-farm research and evaluation to assess the changes.

LWWC, formed in 1991 to serve the common interests of growers in the region. All growers in the region are required to be members of LWWC, and pay a tax of .35 percent of the value of their winegrape gross earnings per year to the commission. There are approximately 650 members with farms ranging from small family farms of five acres to ranches of 9,000 acres, with a median size of about 40 acres.

LWWC began an Integrated Pest Management (IPM) program in 1992. Based on the progress of that program, the Commission was awarded a three-year Biologically Integrated Farming Systems (BIFS) grant from the University of California in 1995. This allowed LWWC to develop its activities, including grower outreach, field implementation, and evaluation, operating on a model featuring grower-driven efforts and collaborative relationships among farmers, scientists, and advisors. As that program generated positive results, the LWWC was awarded additional funding from EPA and other agencies to expand the outreach and impacts.

The IPM program is only one of the commission's activities and priorities, but interviewed growers say that the program has gained importance to them over time. The components of this program include understanding the ecology and dynamics of the crop, and of the pests and natural enemies; developing a monitoring program to assess levels of pests and their natural enemies; establishing an economic threshold for pests; and considering and determining the most appropriate strategies based on the consideration of economics, health, and environmental risks. Outreach activities, such as monthly breakfast meetings with growers, research seminars, and field workshops, helps information exchange and encourages communication among LWWC members.

The LWWC growers have had a range of reactions to the introduction of new integrated practices. Some have been enthusiastic adopters of the biologically integrated practices, and have become strong advocates of the program and educators to other growers. The LWWC members include some organic farmers as well, who have adopted all of the recommended practices, and gone beyond that. On the other hand, some have still been skeptical or resistant to change, especially if they are under economic pressure due to low prices. Nevertheless, growers themselves say there have been significant overall changes in attitude toward these approaches, with more openness and support rather than skepticism.

A grower survey undertaken in 1999, based on self-reporting of 288 growers in LWWC, illustrates some of the results: 76 percent of growers said they reduced their per-acre rates of insecticides; 66 percent reduced their rate of herbicides when spraying for weeds; 46 percent use cover crops; 65 percent monitored for beneficial insects; among other things.

LWWC also undertakes marketing activities. While there has been discussion of creating an "eco-label" for regional wines that reflects the commission's evolving ecological practices, the idea has been put on hold for now, since the commission's work in this area is focused on monitoring practices and applying innovative self-assessment techniques for growers. Marketing efforts include advertising campaigns about Lodi growers and wine, participation in trade shows, industry conferences, media and press kits, public presentations, receptions, special events, membership in wine education associations, and networking services to link growers with market opportunities.

New grants and awards are allowing LWWC to grow the integrated farming program. They have dedicated considerable time and resources to the expansion of their self-assessment program and workbook, which is used for monitoring progress in the adoption of integrated practices. The workbook has been popular among growers and is a model that has been adapted by other winegrowing groups in other regions. Those involved in this effort say they are proud that this program is serving the common economic interests of producers in the region. They also believe that other groups and agribusinesses can follow LWWC's lead in developing biologically integrated approaches through collaborative action.

Information sources for the case study include: Ohmart, Cliff, 1998; Lodi Woodbridge Winegrape Commission's Biologically Integrated Farming System for Winegrapes; LWWC; Lodi, CA, LWWC website; and interviews with Cliff Ohmart, Mark Chandler, Jeff Dlott, and UCSAREP staff.

PROFILE

Lundberg Family Farms

Lundberg Family Farms in Richvale, California produces numerous varieties of rice and rice-based products, and provides about 65 percent of the organic rice grown in the country. The family has been farming rice in California since the 1930s, when Albert and Frances Lundberg and their four children migrated west from Nebraska. Escaping from the "Dust Bowl," they wanted to avoid the serious soil erosion problems they had experienced in the Midwest, and therefore became stewards of the land starting early in their California farming enterprise.

Lundberg Family Farms pioneered organic rice farming in 1969, and continued to develop and extend its ecological approaches over time. The Lundbergs also built a rice processing plant which they have expanded over the years. The company is now fully integrated -- including rice production, processing, packaging, contracting with growers, and marketing a variety of rice and rice-based products.

Four Lundberg brothers and their children currently farm approximately 3,200 acres. About half of their total acreage is organically certified by CCOF. The company currently has the nation's largest brand of organic rice. The other half of the land is 'Nutra Farmed' -- a term coined and patented by the Lundbergs that refers to an integrated farming approach, using a minimum of chemical pesticides and fertilizers. The Lundberg brothers say that in both approaches -- organic and Nutra-Farming -- they are committed to 'sustainable agriculture.' For both,

they use ecologically-oriented practices including cover cropping, crop rotation, water conservation, straw incorporation (and not burning rice residues), and wildlife conservation. In organic fields, they add a few features, such as fallowing the land every 2 or 3 years, allowing it to rest and regenerate, and also using compost or other organic amendments.

Bryce Luncberg explains that the family farms in *both* organic and non-organic ways largely because they respond to diverse demands in the market. In other words, some customers want organically grown produce and will pay the higher price, whereas others do not want to pay the premium. Moreover, organic fields yield approximately half the amount of rice per acre as the Nutra Farmed fields. While the premium they receive for organic rice helps offset those lower yields, the organic systems usually entail higher costs from the fallow periods or other factors. This economic challenge is therefore another reason why they 'Nutra-farm' nearly half of their rice acreage with selective use of chemicals.

The Lundbergs have avoided using 'middle men' in the supply chain to the extent possible. They mill, process, and package their own rice. In recent years, they installed innovative grain coolers for the post-harvest handling process, which enabled them to significantly increase the milling yields. They produce both brown whole grain rice and white (milled) rice of about 12 varieties. They supply not only specialty brown rices and blends, but since the 1980s, they also process, package, and market a wide variety of rice products, including hot brown rice cereal, rice cakes and crackers, dessert pudding, one-step entrees, risotto, rice syrup, and rice flour. Recently, they have also sold some barley, which is mainly used as a cover crop.

The Lundberg family emphasizes producing high quality wholesome products, and maintaining quality standards throughout the entire growing, storage, processing, and handling stages. They have sophisticated and complex storage facilities in order to keep all of the varieties separate and to ensure adequate moisture levels. The company has about 135 employees, most of whom work in the milling and marketing operations.

In addition to processing their own rice, the Lundbergs also buy rice from other growers who together cover about 4,500 acres in Northern California, increasing their capacity to process and sell more rice products. Approximately 25 contracted growers work with the family, ranging in scale from 10 acres to 1,500 acres; these are mostly certified organic. The Lundbergs work closely with their contracted growers to provide information and advice.

Lundberg rice products are sold throughout the U.S., and about 5-10% of their sales are to Canada and Japan. They have also begun to explore market opportunities in Europe. The company markets the products through wholesale companies to an extensive network of natural food businesses, specialty grocery stores, and a few mainstream supermarkets.

They receive a premium price for organic rice, which can be 50% to 80% higher than the price for conventional rice, but the current price for conventional rice is extremely low, so it is hard to make comparisons. Lundberg Family Farms' revenues have grown steadily over time, at a healthy rate of 5 to 8% per year. The family has generally preferred to grow the business by expanding the diversity of their rice-based products and contracting with additional growers, rather than buying more land.

The Lundberg family has continued and expanded the ecological philosophy that was seeded by their father and grandfather two generations ago, by using many practices that are aimed at building their "partnership with nature," and by responding to the social concerns of consumers and neighbors. The Lundberg family's experience also shows how an agricultural business can thrive and prosper by using sustainable methods of agriculture.

Informoation sources include: Interviews with Bryce Lundberg, Lundberg website; analysis of Lundberg Family Farms in NAS, 1989; Alternative Agriculture, National Academy of Science book, and other news clips and experts in this field.

PROFILE

Natural Selection Foods

Natural Selection Foods (NSF) is one of the country's largest growers and processors of packaged specialty and organic salad mixes. NSF also grows and sells other produce, both conventional and organic. The company's organic brand is Earthbound Farm, which began in the mid-1980s as a two-acre organic farm in Carmel, California. Earthbound Farm was founded by Drew and Myra Goodman, who are now president and vice-president of Natural Selection Foods. They have unique story of transformation and remarkable growth.

The Goodmans moved to California from New York in the early 1980s to attend college. Although they did not have previous farming experience, they began raising raspberries and then mixed specialty greens in their backyard garden. They preferred to use organic methods from the start, because they wanted to avoid the use of chemicals. Early on, the couple invented a packaging innovation of bagging prewashed lettuces in plastic bags. The idea stemmed from their home use of convenient zip-top storage bags. Once they started marketing their specialty greens with this kind of package, the products became highly popular with retail buyers. Their business took off with leaps and bounds.

By 1988, the Goodmans employed several people to help them, and established partnerships with salad growers in Southern California. At the same time, they began marketing bagged salads to major mainstream supermarkets. They then bought a 32-acre farm in Watsonville, California, where they planted about 20 varieties of lettuces and greens. In 1992, the company moved their processing and packaging operation to a large production facility in Watsonville. They also opened a farm stand in Carmel Valley. Soon after that, they introduced salad kits, with mixed lettuces, dressing and toppings all in one package.

In 1995, Earthbound Farm entered a partnership with Mission Ranches, a large group of farmers in the Salinas Valley, and formed Natural Selection Foods with 800 organically farmed acres. All of the elements of the operation grew in tandem with increased acreage. By 1998, NSF had 5,800 acres of owned and contracted certified organic farmland dedicated to their product, in California, Arizona, and Mexico. In 1999, NSF merged with Tanimura & Antle (T&A), the largest lettuce grower in the United States. T&A became a one-third partner in NSF and began converting 1,500 acres of farmland into organic production. Earthbound Farm remains the company's leading organic brand.

NSF farms more than 7,000 organic acres today, with more than 2,000 acres in the required three-year transition period from conventional agriculture to organic agriculture. They grow about 85 different fruits and vegetables.

All of Earthbound Farm products and processing are certified organic, and they've been able to adapt these methods to their growing scale of production.

The company's organic approach is described as "nourishing and replenishing soils, protecting water, and honoring the health of those who work the land and customers who will enjoy the harvest." Cover crops, mulch and composting, beneficial insects, careful crop selection, and other organic methods help maintain soil fertility and disease- and pest-resistant plants. Quality assurance and food safety are priorities for Earthbound Farm/NSF, from the farm to the processing facility. A Hazard Analysis Critical Control Point (HACCP) program focuses on employee training and state-of-the-art technologies to maintain safety and quality standards.

The Goodmans still maintain the Earthbound Farm produce stand on their original farm site, selling fresh-picked produce direct to consumers. They've also established an educational plot on the grounds to help children experience agriculture and learn about organic farming.

This transformation took place through innovation, growth, good timing, and strategic partnerships. While Earthbound Farm is undoubtedly a success story in terms of financial success and growth, the enterprise has not been without challenges. Assuring consistent supplies of organic crops, tackling farming problems, and hiring educated and experienced staff have required steady effort and creative strategies. Myra and Drew Goodman credit their success in part to growing their business without training or preconceived ideas about the food industry, allowing them to explore and make decisions with a "beginner's mind."

Today, Earthbound Farm ranks as one of the largest organic foods brands in the world. Though some organic pioneers and consumers fear the "corporatization" of organic farming and hold up NSF as one example of this trend, the company feels it is fulfilling its mission of making organic foods widely available to many people.

Information sources for this case study include: Interviews with Myra and Drew Goodman, Mark Marino, and other staff members; company website; news articles; field tour of Earthbound farm as part of Ecofarm conference; presentation by Rick Antle, conference on Partnerships for Sustainable Agriculture, May 2001; and presentation by Myra and Drew Goodman, Ecofarm conference, Asilomar 2002.

PROFILE

Robert Mondavi Winery

Robert Mondavi Winery is one of the largest exporters of premium California wines, selling to 90 countries. Since its inception in 1965, the company, headquartered in California's Napa Valley, has upheld a land stewardship philosophy based on the convictions of founder Robert Mondavi. Mondavi's early interest in resource stewardship was shaped by the influence of his Italian parents, who instilled in him an appreciation for the land, natural home-grown food, fine wine made from natural processes, and culinary arts, which Mondavi describes in his book *Harvests of Joy*.

Robert Mondavi's winery-owned vineyards total approximately 5,300 acres and are spread throughout several regions of California. In addition, Mondavi buys large amounts of winegrapes from contracted growers throughout California. The company also has joint partnerships in several countries of the world, including Chile, Australia, and Italy, where they both co-own and contract fruit grown by external growers.

Building upon Robert Mondavi's philosophy, the winery developed and articulated a "natural" approach to wine production in the 1970s, with explicit goals of environmental protection, worker health, and enhanced wine quality. The term "natural" in the company's perspective means that they use ecologically-oriented practices at all stages, from soil preparation in the vineyard to bottling practices in the winery.

In the vineyards, for example, they use integrated ecological pest, crop, and soil management methods with selective and minimized use of synthetic chemicals; watershed management; soil and habitat conservation; and minimal tillage. Vineyard managers combine various methods that they judge to be both environmentally and economically sound; they adapt practices to local ecological and climate conditions, rather than using prescribed standardized applications of inputs. Vineyard managers use cover crops and other soil conservation methods such as buffer crops, and they are actively involved in watershed stewardship projects with the community. The company is undertaking a large experiment on the management of wildlife, habitat, and other resources in their Central Coast vineyards. This unusual project entails cooperation with university scientists and state agencies to find potential compatibility between conservation interests and winegrape production.

In the winemaking process and winery operations, Mondavi's practices include the use of native yeast in the fermentation process, energy conservation, and water recycling. Their winemakers support a traditional European approach of bringing out the innate qualities of the grapes using natural ingredients, with minimal interference. In 1994, as part of its environmental efforts, Mondavi created a bottle design free of any metal seal on top. This innovation has been adopted throughout the industry. Mondavi also makes labels from recycled paper and prints them with soy-based inks; uses biodegradable soaps and heat for sterilization; maintains strict standards for use and disposal of oils and solvents.

Robert Mondavi's son Timothy is the winery's managing director today, leading the company's pursuit of natural methods. Tim Mondavi says he believes in continual learning, flexibility, and evolution of ideas as avenues to progress and excellence. According to Tim Mondavi and others in the company, winegrape growing using natural or ecological approaches and minimal chemicals has proven economical without jeopardizing the quality of the product. In fact, some of the company's vineyard production managers believe that this approach actually enhances wine quality and flavor.

The company's transition to these integrated and ecological vineyard practices has not always been easy, since it requires a significant change in attitudes, shifts in costs, and learning new techniques. In addition, convincing growers to consistently adopt Integrated Pest Management (IPM) and other ecological methods can be challenging and requires constraint education. Mondavi's grower relations' managers provide information to growers and strongly encourage that contract growers use natural methods. Although Mondavi works with some organic farmers and embraces some organic methods, they have not yet converted to certified organic methods in their own vineyards, mainly due to economic challenges of weed control. Not all synthetic chemicals have been eliminated yet, though this is a goal of the company.

Robert Mondavi continues to explore new ways of developing environmentally friendly and economically competitive approaches. The company is very open and committed to sharing information with other businesses and the public about these issues and practices. They often hold seminars and educational events not only for their contracted growers, but also for the broader public. Robert Mondavi himself believes strongly in the open exchange of information. He has passed on a philosophy that "what is good for us is good for the industry..." and vice-versa, says DeWitt Garlock, growers relations manager. This form of honest communication and outreach can help broaden and sustain positive socio-economic and environmental outcomes for both current and future generations.

Information sources for this case study include: Interviews with Tim Mondavi, DeWitt Garlock Mitchell Klug, Dan Bosch, Dyson Dimarra, Clay Gregory, Genvieve Janssens, and other staff members; unpublished materials; Mondavi website; discussions with Napa County Resource Conservation District staff, NSWG members; and Robert Mondavi, 1998, Harvest of Joy, Harcourt Brace, New York.

PROFILE

Sherman Thomas Ranch

Sherman Thomas Ranch has recently become recognized in Madera County, California, for its successful transition to biologically integrated and organic farming practices on a 700-acre farm of almonds, pistachios, prunes, and raisin grapes. Under the management of Mike Braga, the manager, the ranch began experimenting during the 1990s with integrated low-chemical-input practices. Since then, about 75 percent of the ranch has been converted to certified organic production. The company is also vertically integrated, operating a dehydrator for processing prunes and raisins, and running a retail produce store.

The ranch's history goes back to the 1930's when it was founded by Sherman Thomas and his family. For many decades, Sherman Thomas owned and conventionally farmed over 30,000 acres that included row crops such as cotton and alfalfa, and tree crops, and pastures for grazing cows, and also had a dairy. During Sherman Thomas' later years, the operation was scaled-down and much of the land was sold, and Mike Braga became the farm manager in 1990. When the elder Thomas passed away in 1995, the remaining 700-acre ranch was passed to the ownership of his son, Vernon Thomas, and is still managed and run by Mike Braga.

The company began to develop integrated pest management practices in the 1980s and early 1990s, and eliminated the use of organophosphate pesticides. During the 1990s, Sherman Thomas ranch became a participant in the Biologically Integrated Farming System (BIOS) program, which was coordinated by the Community Alliance of Family Farmers, and the University of California Sustainable Agriculture Research and Education Program. The BIOS project consisted of on-farm experiments and demonstrations for cover cropping, composting, and reducing pesticide and fertilizer inputs in almond production. Braga was actively involved in trying out BIOS methods, working along with other growers, scientists, and farm advisors.

Since Braga was pleased with the results in the initial BIOS experimental fields on his farm, he continued to expand his land under these integrated practices. Braga became involved in a similar project in prunes, experimenting with biologically integrated methods, and became convinced that the methods paid off. Soon he went beyond that, to develop organic methods in all of his crops. Braga explained that once he had adopted the BIOS practices, "it was easy to eliminate chemicals...and to convert to organic." In fact, his conversion was relatively rapid. Currently, approximately 540 acres are organically certified by CCOF, and the rest is in transition. Braga uses diverse annual cover crops, compost amendments, and no-tillage farming. He emphasizes the use of good sanitation practices to avoid the spread of diseases and insects.

Braga says that he is pleased with the economic results of the organic approach. Although the yields are usually reduced by 25% compared to the conventional approach, they receive a premium (ranging from 25% to 100% higher) for the organic products and spend less on chemicals, so the returns balance out to be similar to or better than conventional. The company has about 8 permanent employees and hires an additional crew of about 25 people during harvest. The organic conversion has not required adding more employees, but the staff has had to learn new approaches.

The company adds value by doing some of their own processing and direct marketing. They own a large dehydrator facility, which enables them to dry their own organic prunes and market the finished product to wholesalers. They dehydrate prunes and raisins for other growers, mainly for custom-order organic raisins for three other farms. However, their own raisin grapes are harvested in the field and allowed to sun-dry naturally in the vineyards. They sell their almonds and pistachios to a certified organic processing company in Fresno.

Sherman Thomas's retail store, called "Valley Pistachio Country Store," sells products that are mostly conventional and locally grown and are purchased largely from wholesalers. They have tried selling some of their own organic produce in the store, but sales have been slow in the Central Valley location, so they sell mostly conventional crops in this small store. They have better results marketing the organic products to areas where there is higher demand. About 75% of total sales is in the United States, and about 25% is exported to Europe.

Sherman Thomas's recent conversion and success in the organic business has become both "a curiosity" and a model for other growers in the county. Braga says that conventional growers frequently come by his farm to ask how to do this: "They often don't think it's possible...and they tend to fear the unknown." But Braga's experience has shown them that it is not only possible but also lucrative. Braga has become a supporter and communicator about organic farming, and is now the president of the local chapter of CCOF. According to a local extension farm advisor, Brent Holtz, "Braga's success has created a following...".

Information sources include interviews with Mike Braga, Brent Holtz, Cindy Lashbrook, and Lonnie Hendricks. Another source of information was an article in California Nuts Magazine, *called "Batting Cleanup," 2002.*

PROFILE

Small Planet Foods

Small Planet Foods (SPF) is one of the largest organic processed foods companies in the United States. Now a subsidiary of General Mills, Small Planet Foods represents, for many, the mainstreaming, growth, and consolidation of the organic foods industry — which has generated mixed reactions in the organic sector.

Small Planet Foods' divisions or brands began as independent companies that each have long-time roots in the organic and natural foods community. They include Cascadian Farm, based in Sedro-Wooley, Washington, a processed and frozen-foods company founded by a pioneer organic farmer, Gene Kahn, currently CEO of Small Planet Foods; and Muir Glen, a California-based company specializing in processed organic tomato products.

Cascadian Farm was founded in 1972 on Gene Kahn's small farm of 22 acres in Washington state, still in operation today. Along with the young natural foods market, Cascadian Farm grew as a processed foods company through the 1970s and 1980s. By the 1990s, investors such as Welch Foods, Inc., and Shamrock Company, helped fuel the company's rapid expansion into mainstream markets. Muir Glen began in 1991 and today remains the industry's leading manufacturer of organic tomato products. In 1998, Small Planet Foods was formed as the umbrella company for both Cascadian Farm and Muir Glen. Fantastic Foods, a maker of processed vegetarian products based in California, was also part of SPF for 3 years, but it was excluded from the partnership in 2000, since its products were not entirely organic.

Under USDA's national organic standards, scheduled for implementation by late 2002, organic processed foods must contain 95 percent certified organic ingredients in order to bear the "organic" label. To meet this requirement, Small Planet Foods currently contracts with about 50 growers who are mostly in the Pacific Northwest, and with 6 to 10 tomato growers in California for the Muir Glen products. These growers range from small organic growers of 10 acres, up to about 2,000-acre operations. Ingredients not grown domestically are sourced internationally; for example, organic sugar, and bananas are imported from Latin America.

As Small Planet Foods, these companies together produce and market approximately 200 processed products, including a wide range of frozen vegetables and frozen fruits, juice concentrates, convenient entrees, frozen desserts, canned tomato products, jams, and sauces. The company's sales grew at a high rate during the late 1990s, up to 20% per year. Annual sales reached $90 million at the end of 2001, and marketing extends throughout the United States, and in Europe and Asia (for 10% of their sales). SPF's stated mission is "to create the world's preeminent organic food company by communicating a powerful vision of the relationship between diet, health, agriculture, and the environment." One of the company's main goals, according to Kahn, is "to become the premier provider of natural and organic products, catching the growing wave of interest in natural foods and a natural way of life."

The companies of Small Planet Foods do not own processing facilities; instead, they contract with about 30 plants. These facilities are also certified organic for processing and packaging methods. Most sales take place through distributors, who in turn sell SPF products to both natural foods and mainstream markets; only about 10 percent is sold directly to retailers. SPF's marketing and packaging strategies position its products to be competitive in both conventional supermarkets and natural foods markets such as Whole Foods Market, and Wild Oats, which sell SPF's brands. Cascadian Farms still runs a small road-side store next to their original 22-acre farm in the foothills of the Cascades.

In 2000, General Mills acquired Small Planet Foods — a noteworthy takeover of an organic business by a transnational food company that received lots of media attention. Since then, the company's sales continue to grow at a fairly high rate of about 10% per year. At the same time, the company buyout has also been criticized, especially from smaller scale organic pioneers, for this 'corporatization' of the organic industry. The takeover has also generated questions about the sustainability and social responsibility of the situation, since many other organic businesses face difficulties to survive when trying to compete against such large companies. However, Gene Kahn and his staff, and others in the business, believe this ownership by General Mills enables SPF's organic products to be marketed more economically and purchased by many more mainstream consumers throughout the United States and the world.

Information sources for this case study include: Interviews with Gene Kahn, Clark Driftmeir, Craig Weakley, Steven Crider, Lawrence Tsai, Lisa Bell, and other staff members; company website; news articles such as Deann Glamser, 1998, "Organic Growth," Your Turn, Winter (trade magazine), and other news articles; and presentations by Craig Weakley at Ecological Farming conference and "Partnerships for Sustaining California Agriculture" conference.

REFERENCES

Altieri, Miguel. 1987. *Agroecology: The Scientific Basis of Sustainable Agriculture*. Westview, Boulder, CO.

Altieri, Miguel. 1992. "Agroecological Foundations of Alternative Agriculture in California." *Agriculture, Ecosystems, and Environment* 39: 23-53.

Ames, Paul. 2000a. "Organic farming thriving in Europe." Associated Press, *Wichita Eagle*, January 2.

Ames, Paul. 2000b. "Organic farming a growing trend." *Orange County Register* (Santa Ana, CA), January 7.

Arnold, Matthew, and Robert Day. 1998. *The Next Bottom Line: Making Sustainable Development Tangible*. World Resources Institute, Washington, DC.

Auburn, Jill. 1994. "Society Pressures Farmers to Adopt More Sustainable Systems." *California Agriculture*. 48(5): 7, pp. 9-10.

Batie, Sandra, and D.B. Taylor. 1989. "Widespread Adoption of Non-Conventional Agriculture: Profitability and Impacts." *American Journal of Alternative Agriculture*, 4(3-4): 128-134.

Benbrook, Charles. 1996. *Pest Management at the Crossroads*. Consumers Union, Yonkers, New York.

Beus, C. E. and R.E. Dunlap. 1990. "Conventional Versus AlternativeAgriculture: The Paradigmatic Roots of the Debate." *Rural Sociology*, 55(4): 590-616.

Bourne, Joel. 1999. "The Organic Revolution." *Audubon Magazine*, March-April.

Brazil, Eric. 2001. "Organic Farming Sprouts Businesses." *The Chronicle*, San Francisco, April 22. A25.

Buck, David, and Christina Getz, and Julie Guthman. 1996. *Consolidating the Commodity Chain: Organic Farming and Agribusiness in Northern California*. Development Report, Food First, Oakland, CA

Carson, Rachel. 1962. *Silent Spring*. Houghton Mifflen, New York.

Collins, Keith. 2000. "Outlook for the Farm Economy in 2000." *Agricultural Outlook*, April. Pp. 2-4

Condor, Bob. 2000. "Organic Market Growing at Record Speed." *Chicago Tribune*, September 17.

Conway, Gordon, 1987. "The Properties of Agroecosystems." *Agricultural Systems*, Vol 24: 95-117.

Conway, Gordon. 1998, *The Doubly Green Revolution: Food for All in the 21st Century*. Cornell University Press, Ithaca, New York.

Conway, Gordon, and Jules Pretty. 1991. *Unwelcome Harvest: Agriculture and Pollution*. Earthscan Publications, London.

Corselius, Kristen, Suzanne Wisniewski, and Mark Ritchie. 2001. *Sustainable Agriculture: Making Money, Making Sense*. Institute for Agriculture and Trade Policy, Minneapolis, MN.

Dmitri, Carolyn, and Nessa Richmond. 2000. "Organic Foods: Niche Marketers Venture into the Mainstream." *Agricultural Outlook* (Economic Research Service, USDA) June-July: 11-15.

Doering, O. 1991. "U.S. Federal Policies as Incentives or Disincentives to Ecologically Sustainable Agricultural Systems." Staff paper #91-14. Department of Agricultural Economics, Purdue University, West Lafayette.

Douglass, Gordon K. 1984. "The Meanings of Agricultural Sustainability," in G.K. Douglass, ed, *Agricultural Sustainability in a Changing World Order*, pp. 3-29. Westview Press, Boulder, CO.

Dunn, J. 1995. *Organic food and fiber: An Analysis of 1994 certified production in the United States*. USDA Agricultural Marketing Service, Washington, DC.

Ellis, William. 1991. "Harvest of Change." *National Geographic*, February

Ervin, David, and Sandra Batie. 2000. *Transgenic Crops: An Environmental Assessment*. Henry Wallace Center for Agricultural and Environmental Policy at Winrock International. Washington, DC.

Economic and Social Research Programme, 1999. *The politics of GM food: Risk, science and public trust*. ESRC, Centre for the Study of Environmental Change, Lancaster University, UK.

Fabricant, F. 1995. "Organic Foods Go Mainstream." *New York Times*. November 6, 1999, p. C3.

General Accounting Office. 2001. Agricultural Pesticides: Management Improvements Needed to Further Promote Integrated Pest Management (GAO-01-815). General Accounting Office, Washington, DC.

Gershuny, Grace. 2000. "Do we continue to support the Organic Foods Protection Act?" *Organic Farmer*, pp. 31-33.

Gilmore, John. 1999. "U.S. Organics 1998." DATAMONITOR, New York, NY.

Gliessman, Steven R. 1992. "What are the Indicators of Sustainability in Farming Systems?" in *Organic '92: Proceedings of the Organic Farming Symposium*, UC Division of Agriculture and Natural Resources, Publication 3356, pp. 46-48. Davis, CA.

Greene, Catherine. 2000a. "U.S. Organic Agriculture Gaining Ground." *Agricultural Outlook*, Economic Research Service, USDA. April issue, pp. 9-14.

Greene, Catherine. 2000b. "Adoption of Organic Farming Systems: Progress in the 21st Century?" Economic Research Service, USDA, Paper from the Organic Agriculture Symposium in the American Association for Advancement of Sciences conference.

Guthman, Julie. 1999. "Raising Organic: An agroecological assessment of grower practices in California." draft paper, Department of Geography, University of California, Berkeley.

Hartman Group. 1996. *The Evolving Organic Marketplace*. Summer. Hartman Group, Bellevue, WA.

Hartman Group. 1999. *Food and the Environment: A Consumers' Perspective, Phase 3*. The Hartman Group, Bellevue, WA.

Hartman Group. 2000. *Organic Consumer Profile*. The Hartman Group, Bellevue, WA.

Hartman Group. 2000. "Organic Lifestyle Shopper Study - Summary," on website www.hartman-group.com/organicstudy.html, The Hartman Group, Bellevue, WA.

Harvard Business Review. 1994. "The Challenge of Going Green." *Harvard Business Review*, July-August, pp. 37-50.

Hawken, Paul. 1994. *The Ecology of Commerce: A Declaration of Sustainability*. Harper, New York.

Hawken, Paul, Amory Lovins, and H.J. Lovins. 1999. *Natural Capitalism: Creating the Next Industrial Revolution*. Little Brown and Company, New York.

Heffernan, William. 1999. *Consolidation in the Food and Agriculture System*. Report to the National Farmers Union. (Rural Sociology Dept, University of Missouri, Colombia).

Hendrickson, Mary. 2000. *Consolidation in Food Retailing and Dairy: Implications for Farmers and Consumers in a Global Food System*. National Farmers Union. (Rural Sociology Dept, University of Missouri, Colombia.)

Hesterman, Oran B and T.L. Thorburn. 1994. "A Comprehensive Approach to Sustainable Agriculture." *Journal of Production Agriculture*. 7(1): 132-134.

Ikerd, John. 2000. "Organic Agriculture Faces the Specialization of Production Systems: Specialized Systems and the Economical Stakes," Working paper, Agricultural Economics Department, University of Missouri.

International Trade Centre (ITC). 1999. *Organic Food and Beverages: World Supply and Major European Markets*. International Trade Centre, UN Center on Trade and Development, Geneva.

Johnson, Douglas. 1996. "Green Business: Perspectives from Management and Business Ethics." *Society and Natural Resources*, Vol 11: pp. 259-266.

Jolly, Desmond. 1991. "Differences Between Buyers and Nonbuyers of Organic Product and Willingness to Pay Organic Price Premiums." *Journal of Agribusiness*. 9(1): 91-111.

Kirschenman, Frederick, G, Kahn, and A. Ferguson. 1993. "Towards a Sustainable Organic Food Marketing System." *Organic Farmer* 4(2): 16.

Klinkenborg, Verlyn. 1995. "A Farming Revolution: Sustainable Agriculture." *National Geographic*, December: 65-88.

Klonsky, Karen and P. Livingston. 1994. "Alternative Systems Aim to Reduce Inputs, Maintain Profits." *California Agriculture* 48(9): 34-42.

Klonsky, Karen, and L. Tourte. 1993. "Statistical Review of California's Organic Agriculture." California Department of Food and Agriculture. Organic Program, Sacramento, CA.

Lampe, Frank. 2000. "Homegrown Rollup Creates $50 Million Foods Player," *Natural Business*. October issue, p 1.

Liebman, James. 1997. *Rising Toxic Tide: Pesticide Use in California 1991-95*. Pesticide Action Network, San Francisco, CA.

Lipson, Elaine. 2000. "Conagra Swallows Lightlife Foods." *Natural Business*, August issue, page 1.

Lipson, Elaine. 1997. Organic standards (?), New Foods Merchandiser. (PLEASE INSERT CORRECT TITLE AND DETAILS ON VOLUME & PAGE)

Lipson, Mark. 1997. *Searching for the "O" Word: Analyzing the USDA Current Research and Information System for Pertinence to Organic Farming*. Organic Farming Research Foundation, Santa Cruz, CA.

Mayer, Steven. 1998. "Organic." *The Bakersfield Californian*, April 25, 1998.

Mergentime, K, and M. Emerick. 1996. "Widening market carries organic sales to $2.8 Billion in 1995." *Natural Foods Merchandiser*, New Hope Communications, Boulder, CO.

Myers, Steve, and Somlynn Rorie. 2000. "Facts and Stats: The Year in Review." *Organic and Natural News*, December Issue.

National Research Council. 1989. *Alternative Agriculture*. National Academy Press, Washington, DC.

National Research Council. 1991. *Sustainable Agriculture Research and Education in the Field: A Proceedings*. National Academy Press, Washington, DC.

National Research Council. 1996. *Ecologically-Based Pest Management: New Solutions for a New Century*. National Academy Press, Washington, DC.

National Research Council. 2000. *The Future Role of Pesticides in U.S. Agriculture*. National Academy Press, Washington, DC.

Natural Foods Merchandiser. 1997-2000, current and back issues of journal on website, www.healthwellexchange.com/news.cfm

Onstad, Eric. 1999. "Tasty earnings fuel European Organic Sector." *ABC news*. November 23.

Organic Trade Association. 2000. "Food Facts, and Environmental Facts," from Website of OTA www.ota.com.

Perkins, John. 1982. *Insects, Experts and the Insecticide Crisis*. Plenum Press, New York.

Pimentel, David and H. Lehman, eds. 1993. *The Pesticide Question: Environment, Economics, and Ethics*. Chapman and Hall, New York.

President's Council on Sustainable Development. 2000. *Sustainable Agriculture - Policy Recommendations*, President's Council on Sustainable Development, The White House.

Pretty, Jules. 1995. *Regenerating Agriculture: Policies and Practices for Sustainability and Self Reliance*. Earthscan Publications, London.

RAFI. 2000. "The Seed Giants – Who Owns Whom?" Rural Advancement Foundation International website, www.rafi.org

Reisner, Mark. 1993. *Cadillac Desert*. Penguin Inc, New York.

Richmond, Nessa. 1998. *The Natural Foods Market: A National Survey of Strategies for Growth*. Policy Studies Report Number 12, The Henry Wallace Institute for Alternative Agriculture, Washington, DC.

Rissler, Jane, and Margaret Mellon. 1996. *Ecological Risks of Engineered Crops*. MIT Press, Boston.

Roberts, Paul. 1998. "Growth Industry," *Approach* (Reno Airlines) pp. 36-42.

SARE. 1998. *Ten years of SARE: A Decade of Programs, Partnerships and Progress in Sustainable Agriculture Research and Education*. Sustainable Agriculture Research and Education Program, U.S. Department of Agriculture, Washington, DC.

SARE Western Region. 2000. *Sustainable Agriculture: Continuing to Grow. A Proceedings of the Farming and Ranching for Profit, Stewardship and Community*. Conference, March. Western Region Sustainable Agriculture Research and Education Program, and Sustainable Northwest, Portland.

SARE. 2001. *The New American Farmer: Profiles of Innovation*. Sustainable Agriculture Research and Education Program, U.S. Department of Agriculture, Washington DC.

Schaller, Neill. 1993. "Farm Policies and the Sustainability of Agriculture: Rethinking the Connections." Henry Wallace Institute for Alternative Agriculture, Winrock International, Washington, DC.

Sooby, Jane. 2001. *State of the States: Organic Farming Research in Land Grant Universities*. Organic Farming Research Foundation, Santa Cruz, CA.

Swezey, Sean, and Janet Broome. 2000. "Growth predicted in biologically integrated and organic farming." *California Agriculture*, 54(4): 26-35.

Thrupp, Lori Ann. 1996. *New Partnerships for Sustainable Agriculture*. World Resources Institute, Washington, DC.

Thrupp, Lori Ann. 1998. *Cultivating Diversity: Agrobiodiversity and Food Security*. World Resources Institute, Washington, DC.

Thrupp, Lori Ann. 1999. *Roots of Change: The Sprouting of Sustainable Agriculture in California*. Unpublished report for the Funders Agricultural Working Group, Clarence Heller Foundation, San Francisco, CA.

UCSAREP. 2000. "What is Sustainable Agriculture?" University of California Sustainable Agriculture Research and Education Program, information on website www.ucsarep.ucdavis.edu, University of California.

Uhland, Vicky. 2000. "Organic Farmers Struggle Despite Rising Sales." *Natural Business*. December issue, p. 1.

United Nations Development Programme. 1995. *Agroecology: Creating the Synergism for a Sustainable Agriculture*. United Nations Development Programme, New York.

United Nations Development Programme. 1992. *Benefits of Diversity*. United Nations Development Programme, New York.

US Environmental Protection Agency. 1998. *Food Production and Environmental Stewardship: Examples of How Food companies Work with Growers*. Policy Planning and Evaluation Report 2128, U.S. Environmental Protection Agency, Washington, DC.

Van der Harst, Tatiana, and Laura Gabel Scandurra. 1997. "Dutch Organic Food Market Offer All Natural Potential for U.S. Firms." *Ag Exporter*, pp. 10-17.

Walz, Erica. 1999. *Third Biennial National Organic Farmers' Survey*. Organic Farming Research Foundation, Santa Cruz, CA.

White, Heather. 2000a. "Kellogg Acquires Kashi Co." *Natural Business*, August issue, page 1.

White, Heather. 2000b. "Nanosecond Consolidation Rocks the Industry." *Natural Business*, March issue, page 1.

Willer, Helga, and Minou Yussefi. 2001. *Organic Agriculture Worldwide 2001: Statistics and Future Prospects*. International Federation of Organic Agriculture Movements (IFOAM), Durkheim, Germany.

Young, Douglas. 1989. "Policy Barriers to Sustainable Agriculture." *American Journal of Alternative Agriculture*, 4 (3-4): 135-141.

Youngberg, Garth, Neill Schaller, and Kathleen Merrigan. 1993. "The Sustainable Agriculture Policy Agenda in the United States: Politics and Prospects," in P Allen ed. *Food for the Future: Conditions and Contradictions of Sustainability*, pp. 295-317. John Wiley and Sons, Inc.

APPENDIX 1

PEOPLE INTERVIEWED AND CONSULTED FOR THE STUDY

Information for the case studies was obtained through structured interviews with the directors and staff of each company, and with external informants who are familiar with the cases. Questions were posed about: General characteristics of the case/company; Agricultural production practices; Reasons for and results of using "green" practices; Food processing practices; Marketing practices; Economics; Information sources and linkages in the supply chain; Barriers and challenges; Future plans and prospects.

Interviews were undertaken with directors and staff of 12 case studies, which included the following: Del Cabo, Durst Farms, Fetzer Vineyards, Frog's Leap, Full Belly Farm, Lagier Ranches, Lodi Woodbridge Winegrape Commission, Lundberg Family Farms, Natural Selection Foods, Robert Mondavi Winery, Sherman Thomas Ranch, Small Planet Foods. (The names of interviewees are noted in footnotes to the profiles in Part II.) In most of the cases, the farms/sites of the companies were also visited for this analysis.

Additional experts and analysts were informally interviewed, consulted, or provided insights during oral presentations. They provided information about historical and policy aspects, perceptions of progress and challenges, and other issues related to sustainable agriculture, marketing, and food systems. These individuals include the following:

Michael Abelman, Fairview Gardens
Miguel Altieri, U.C. Berkeley, Agroecology
Rick Antle, Tanimura and Antle company
Jill Auburn, SARE, U.S. Department of Agriculture
Walt Bentley, U.C. Extension, pest management specialist
Jenny Broome, U.C. Sustainable Agriculture Research & Education Program
Amigo Bob Cantisano, Organic consultant
Stacey Clary, California Sustainable Agriculture Working Group
Jeff Dlott, Realtoolbox consulting
Volker Eisle, winegrape grower, Napa Valley
Isao Fujimoto, California Institute of Rural Studies
Catherine Greene, ERS, U.S. Department of Agriculture
John Ikerd, University of Missouri
Bruce Jennings, Policy expert, Sacramento, CA
Desmond Jolly, Small Farm Center, U.C. Davis
Fred Kirchenmann, Grower, and Leopold Center, Iowa State University
Sibella Kraus, formerly Community Alliance of Family Farmers
Bill Liebhardt, U.C. Davis
Ralph Lightstone, policy expert, Sacramento, CA
Mark Lipson, Organic Farming Research Foundation
Craig McNamara, Sierra Orchards
Monica Moore, Pesticide Action Network
Bu Nygrens and Mary Jane Evans, Veritable Vegetables
Stephen Pavich, Pavich Family Farms
Peter Price, legal and policy expert/advocate
Mark Ritchie, Institute for Agriculture Trade and Policy
Walter Rob, Whole Foods
Karen Ross, California Association of Winegrape Growers
Bob Scowcroft, Organic Farming Research Foundation
Sean Swezey, U.C. Santa Cruz
Stephen Temple, U.C. Davis, Integrated Farming Program
Alice Waters, Chez Panisse
Warren Weber, Grower, Bolinas, CA
Frank Zalom, U.C. Integrated Pest Management program